The Theory of Stationary Space

THE THEORY OF STATIONARY SPACE

Mark Meek

iUniverse, Inc.

New York Lincoln Shanghai

The Theory of Stationary Space

iUniverse, Inc.

For information address:
iUniverse, Inc.
2021 Pine Lake Road, Suite 100
Lincoln, NE 68512
www.iuniverse.com

ISBN: 0-595-33909-3

Printed in the United States of America

I would like to dedicate this book to everyone that is going to join me in the Rapture.

Contents

▼

INTRODUCTION

Human beings have certainly made great strides over the last hundred years or so concerning our understanding of the nature of the universe. Einstein's two theories of relativity as well as the dissection of the atom into it's component particles were nothing less than revolutionary. The discovery and measurement of the expanding universe and Quantum Mechanics were just as revolutionary. Maybe we are on our way to a point where we could know just about everything.

However, everything is not quite well. There are some things missing. Something is wrong somewhere. It seemed to me that no matter how much we learn about the nature of the universe or how successful our theories seem to be there are several major unanswered questions about the nature of the cosmos and the world around us. There are questions that physicists and cosmologists often seem to skip around when putting theories about the nature of the universe together. The fact that we know so much makes the fact that we cannot solve these great mysteries all the more puzzling.

First of all, what is time? I mean from a physics point of view. No one seems to know and even Einstein did not try to answer it except to describe it as a function of the speed of light.

That is the next mystery that troubled me. Why can we measure the speed of light with great precision but no one seems to know exactly why this is the speed of light. There seems to be no physical reason for this to be the speed of light instead of some other speed. This just did not seem to be right to me. The speed of light is the basis of Einstein's Special Theory of Relativity. If the speed of light is so important then there should be obvious physical reasons why it was the speed of light.

The most famous formula of the Twentieth Century is certainly Einstein's E=MC squared. This is the formula for the conversion of matter and energy. The

C is for "constant", which is the speed of light because in the Special Theory of relativity, the speed of light is always constant. In this formula, the speed of light is multiplied by itself, or squared.

But why does E=MC squared? Why does the speed of light enter into a formula for the conversion of matter and energy and why is it squared in the formula? Einstein pointed this out but did not explain why. We really should be able to answer this but we cannot.

The next great mystery is why there is two electrical charges in the universe, negative and positive. Why are there two and only two and from where do these charges originate? Yes, I know that an electron has a negative charge but why is that so? Why are we not able to answer this?

What about mass? Why do some things in the universe have mass, such as a rock, but some do not, like light and space. Simply adapting an attitude of "that's just the way it is" did not seem to me to be good enough.

How about relativity? Einstein told us that the speed of light in a vacuum is absolutely fixed, time stretches out and the mass of an object increases while the object seems to contract in length as it approaches this speed. But why, why, why is this so? For another thing, why in the world should the rate of the passage of time be affected by gravity?

Going all the way back to Isaac Newton's Laws, why does the Law of Inertia apply? That is, why would a body in motion remain in motion until affected by an outside force? We still have not answered that question. As for another of Newton's Laws, why does every action result in an equal but opposite reaction? We know the what but not the why.

These great mysteries kind of dampened, if not ruined, our models of the universe as far as I was concerned. Our inability to provide answers to these questions made our knowledge of the way things are appear like a piece of Swiss cheese full of gaping holes. It seemed as if we knew quite a bit about what was going on in the universe but did not know why.

These are not arcane questions at all but mysteries concerning that which is most primal to our lives and the world around us. Time is literally what life is made of. We are put on this earth and are given a certain amount of time here. The speed of light is the focal point of the Special Theory of Relativity. According to Einstein only the speed of light is fixed, all else is relative. The two opposite electrical charges make atoms possible and govern their behavior as well as that of all electromagnetic radiation, including light. We know that matter has mass but we are not really sure what matter really is, what causes it to exist, or why some things do not have mass (like light and space itself). Einstein's Special Theory of

Relativity is the essence of brilliance but it still does not tell us why nothing can ever travel faster than the speed of light, why time ceases to exist at this speed and progressively slows down while approaching it and why an object would have infinite mass while traveling at this speed. Why does Newton's Laws of inertia and opposite reactions always apply?

These are basic and primal mysteries yet modern physics or cosmology has not really answered any of these questions. No matter how much else is learned, can we really consider physics as a success until these questions are answered and the great mysteries solved?

As it turns out, there is a simple model of the universe that answers all of the above and the model fits with what is known about the universe and much of it is, in fact an arrangement of some of the latest theories. This simple model I have named "The Theory of Stationary Space". It answers so many primal, unanswered questions that it cannot possibly be wrong.

My point of view is that we cannot consider ourselves as successful at understanding the basic nature of our universe if we cannot solve the basic mysteries listed above and if we have to look at things very differently in order to find solutions, then so be it.

CHAPTER 1

▼

THE STATE OF THE UNIVERSE

Our modern view of the universe began when Copernicus dared to put forth the view that the sun, rather than the earth, was at the center of the Solar System. Galileo confirmed with his telescope that this was indeed the case. Isaac Newton defined gravity as the basic force that operates the universe. Newton also established the basic laws of motion. A body in motion tended to stay in motion until acted upon by an outside force. The inertia of a body was equal to it's mass multiplied by it's velocity. Every reaction resulted in an equal and opposite reaction.

Next came Johannes Kepler, who determined that the orbits of the planets around the sun were not circles but ellipses. An ellipse is like a circle but has two foci instead of one. An imaginary line from the planet to the sun swept over equal distances in equal periods of time. Thus, a planet moved faster in it's orbit when it was closer to the sun and slower when it was further away. Planets further from the sun revolved around the sun slower than the closer planets.

A Danish astronomer named Ole Romer attempted to measure the speed of light in 1676 by making use of eclipses of Jupiter's moons. He came up with a figure of 140,000 miles per second. This is less than the modern established figure of 186,282 miles per second but was not too bad for a first try in 1676. The modern figure was arrived at by the well-known Michelson-Morley experiment.

It was discovered that atoms were not the most fundamental particles at all, as had long been believed. Atoms turned out to be composed of still smaller pieces. The inside of the atom was governed by the same electric charges involved in electric currents. The electrons, with a negative charge, were in orbit around protons, with positive charges. There were also neutrons, with no electric charge, in the atomic nucleus.

Knowledge of physics and astronomy continued to build until in 1905, a theorist with whom we are all familiar named Albert Einstein introduced his "Special Theory of Relativity". This famous theory concerned the speed of light. Einstein postulated that the speed of light in a vacuum was absolute and invariable regardless of the relative velocity of the source of light and would be perceived the same no matter how fast an observer was traveling.

This was a radical idea in so-called "classic" or "Newtonian" physics in which one would logically assume that if someone fired an arrow straight ahead from a moving train that the velocity of the train would be added to that of the arrow.

The Special Theory of Relativity also held that time is not absolute but would seem to slow down if one were to travel close to the speed of light. Einstein used the famous paradox of the twins to illustrate the point. If one twin left earth on a spacecraft traveling at close to the speed of light, he would find his twin much older than he when he returned because time would be progressing much more slowly for the astronaut than for his twin on earth.

Distance too, actually shortens when traveling close to the speed of light. However, nothing can ever exceed the speed of light because an object becomes heavier when traveling at very high speeds and will have infinite mass when traveling at the speed of light. An infinite force would then be required to accelerate it to a higher speed, which is naturally impossible. It seemed really bizarre and mind-bending in 1905 but has been conclusively proven by observation. Einstein pointed out that there are four dimensions, three of space and one of time.

Einstein also pointed out that the mass in matter and energy are actually interchangeable. This was really mind-bending and led to nuclear power and bombs. The exact formula for the conversion was E=MC squared. Energy equals mass times the speed of light squared. Since the speed of light was very, very fast, this meant that a small amount of mass contained a very great amount of energy.

In 1915, Einstein came up with more relativity. This was the so-called "General Theory of Relativity". The theory was developed to bring Newton's old laws of gravity into harmony with the Special Theory of Relativity. This theory concerned space and how it was warped by the presence of mass, such as the sun or a planet. The result of this warp in space is what we know as gravity.

It is not so much that gravity is a force that draws objects together but rather that a mass such as a planet curves space like a bowling ball placed on a mattress and a ball rolling by will fall into "orbit" around the bowling ball just as the moon orbits the earth.

When you throw a ball in a straight line and it curves downward to meet the earth, it is not that a force called gravity is pulling the ball earthward, but that space itself is curved by the earth's mass and the ball travels in a straight line through the curved space to meet the earth. An important part of the General Relativity Theory was that gravity would bend light if it did in fact curve space itself and this has been proven by observation.

Einstein's equivalency principle is that the effects of gravitation and acceleration are the same. He concluded that an observer in free fall cannot tell by his senses that gravity is present. Therefore, Einstein reasoned, gravity must be a part of the space-time through which the observer is falling and not an outside force.

Einstein's theories explained an anomaly in the orbit of Mercury around the sun. The perihelion (closest point to the sun) of Mercury's orbit moved forward in the orbit a little bit each time around the sun. This was actually caused by the effective shortening of distances due to the warp in space itself caused by the great mass of the sun.

Progress continued to be made in our understanding of the makeup of the universe in both cosmology, the field of the very large, and in Quantum Mechanics, the field of the very small. The ultimate in what we are searching for is a so-called theory of everything. Einstein's dream in his later years was to find a way to unify gravity and electromagnetism.

The theory of everything would not necessarily be a simple formula that explained everything that happened but rather a base that everything could be structured on. As one writer called it, "The Universe on a T-shirt" that is, everything reduced to a formula simple enough to be worn on a T-shirt.

In Quantum Mechanics, we now know that the protons and neutrons in atoms are made of three quarks each. Electrons are known to be very tiny point particles with no discernable internal structure and are in a class of particles known as leptons. Physicists have found it useful to assign quarks imaginary colors. This theory is known as Quark Chromodynamics and a transformation can change the color of one quark at the expense of another but the two changes cancel out and the quark remains colorless.

When Quark Chromodynamics is joined to a combination of the electromagnetic force and weak nuclear force (that is linked with radioactivity), it produces a successful theory that we call the Standard Model. Of course, we know that a quark cannot really have any visible color simply because it is much smaller than the wavelength of visible light.

In 1970 a radical new theory came along, known as String Theory or Superstrings. I have wondered what Einstein would have thought of String Theory if he were alive. The basis of String Theory is that there is more than the three dimensions that we see and the most fundamental particles, such as the quarks and leptons, are not particles at all but tiny strings that may be of great length despite being of very tiny cross-section.

In 1919, after Einstein had pointed out that we inhabit a universe of four dimensions-three of space and one of time, a mathematician by the name of Theodor Kaluza proposed that the addition of a fourth spatial dimension may allow the linking of General Relativity and Electromagnetic Theory. In 1926, a Swedish physicist named Oskar Klein found a way to incorporate Kaluza's idea into quantum theory. Ever since then, theories involving dimensions that we cannot perceive with our senses have been known as Kaluza-Klein theories. Those dimensions that we cannot see are known as hyperspace and sometimes as Grassmann dimensions after a physicist of that name.

String Theory has a lot going for it. A major problem in attempts to unify Quantum Mechanics and gravity is the appearance of infinite values in mathematical solutions. But replacing the standard point particles, such as quarks and leptons, with strings does away with the problem. Mathematically, string theory makes a lot of sense even though it always requires that there are more dimensions then we can see. So-called Heterotic String Theory combined two other string theories, one with open-ended strings requiring twenty-six dimensions and the other with closed-ended strings requiring ten dimensions.

In January of 1985, the idea of compactification was introduced. This was the possible curling up of extra dimensions into tiny structures that are invisible to us. The mathematical structures that are supposedly needed for the compactification of heterotic strings are known as Calabi-Yau spaces.

In further theoretical developments, Twistor Theory, by Oxford mathematician Roger Penrose, is an effort to unite quantum theory and relativity. Twistors make up both the points that define space and the particles that inhabit it. Some variations of String Theory call for strings to create their own space-time rather than operating in a background space-time.

Many people today have read "The Universe in a Nutshell" by Stephen Hawking or "The Whole Shebang" by Tim Ferris and see our universe as a membrane or "brane" for short of four dimensions, three of space and one of time, existing in higher dimensional space. This would be much as a two-dimensional piece of newspaper might be floating around in our three-dimensional space.

Some people believe that matter, as well as light, are represented by strings fastened to the brane. But that gravitons, particles of gravity, are looped string and thus are not fastened to the brane. This would mean that light cannot leave the brane (or universe) but gravity can. So, we could never visually see other branes (universes) beyond our own but could possibly detect their gravity and it also implies that the reason our universe is expanding may be because gravity is leaking out of it. It is considered as possible that the Big Bang, the primeval explosion that started our universe, could be the result of our brane colliding with another brane.

Finally, we come to M-Theory. I take the M to stand for "membrane", although others take it to stand for "matrix". M-Theory revolves around two-dimensional membranes or sheets rather than one-dimensional strings. Some versions of M-Theory use waves instead of membranes. M-Theory was revived in the mid-90s after having been abandoned earlier. It is a mathematical complex to describe all forces and particles. There have been at least five variations on M-Theory.

CHAPTER 2

▼

TIME

One thing that still mystifies us is time. What actually is time from a physical point of view? Physics tends to skip around that question, just taking it for a given. What we perceive as time is actually not time at all but rather motion. All of the methods we have for measuring time are actually measuring motion instead of time. It is true that in our everyday world, motion is a fairly close approximation of time. But that still leaves us with the question, what exactly is time? Even Albert Einstein only pointed out that it was a dimension as were the three spatial dimensions.

Have you ever gone to a library and tried to find a book about what time is? There was one book, but it turned out to be a history of clocks. There was another, but it was about cultural perceptions of time. There was still another book, but it was about the biological rythms and clocks in living things. Yet another was about high precision measurement atomic clocks. There is the popular magazine called "Time", but it is a news magazine about world events happening in time.

With all of our knowledge about physics and the universe, why can we not explain something as universal and primal as time? It governs everything we do but we do not really know what it is. By that I mean what time is from a physical point of view. We know that time is a dimension. But why is it that we can move about at will in the other three dimensions, while in the time dimension we can only move forward at a fixed rate and can never go backward or stand still? Have you ever stopped to ponder how bizarre this is?

What makes it really bizarre is that no one seems able to explain the physics of it with much satisfaction other than defining time as "that which goes by one second per second". Why can we move around in space so easily but not in time?

In the opinion of this writer, modern physics cannot really be considered as a success in explaining the universe until it can provide a satisfactory answer to the mystery of time. One of the objects of this book is to change that. If there is one thing I must do in my life it is to try to explain this ancient mystery that is just about as baffling today as it ever was.

There are two basic views of the passage of time in the universe. The first is called Determinism, in which the past leads to the present, which leads to the future. This is primarily a cause and effect scenario that perceives time as motion and in which the present is all that actually exists at any given time. The second view is the so-called Block Universe, in which past, present and, future all exist at once.

TIME DOES NOT REALLY EXIST

My conclusion is that we cannot nail down what time actually is because it does not really exist. It is only something that is perceived by living things. It is our capacity for memory that causes us to believe time is something when it really is not.

There is only one explanation for time that I can imagine that is plausible and makes sense. There are four dimensions. The difference is that contrary to popular belief, there is not three dimensions of space and one of time. That is what our memory causes us to perceive. There are four dimensions of space. What we know as matter, the world and universe around us, consists of strings aligned mostly in one direction in four-dimensional space. That direction is what we perceive as time. The other three dimensions we perceive as space.

We will introduce postulates periodically throughout this book.

POSTULATE 1—THERE ARE FOUR SPATIAL DIMENSIONS. MATTER IS COMPOSED OF VERY LONG, ONE-DIMENSIONAL STRINGS ALIGNED PRIMARILY IN ONE DIRECTION IN THE FOUR DIMENSIONAL SPACE. IT IS THIS DIMENSION THAT WE PERCEIVE AS TIME. THE OTHER THREE DIMENSIONS WE EXPERIENCE AS SPACE.

String Theory is basically correct. It has too much going for it not to be, although there have been many variations of the theory. Each fundamental particle in the universe, primarily quarks and electrons, is really a one-dimensional string. However, if the strings were aligned primarily in one direction like parallel lines or the stripes on an American flag or a striped shirt, this would also explain what time is.

If the strings are going all over the place at different angles, time could not be explained in this way and would require a different explanation. If the fundamental particles are really strings, although they always appear to us as particles, it means that there must exist at least one other dimension that we cannot see. Since we, in our three dimensions, always see strings as particles and not as strings and, that the amount of motion relative to the total potential motion on a large scale is very low, we can safely conclude that the strings are all aligned primarily in one dimension in space. The bonus is that when we accept this definition of time, the answers to so many other mysteries about the universe just seem to fall into place as well. I consider that as proof of this theory. Also, why would gravity have an effect on time, as it does on space, unless time is also space?

We should not be too surprised that we have been wrong in our perception of what time is. After all, until Einstein's General Theory of Relativity explained gravity, another primal entity, as not a force at all but the warp in the fabric of space caused by mass such as the earth, we had incorrectly believed gravity to be the basic force operating the universe.

The infinitesimal points that students of geometry and calculus know that lines and curves are composed of are also a physical model of the universe that we see. An electron is not the same as it was a moment ago. The electron we are looking at is another point on the same string. Our universe or, our brane, is the four dimensions that we can relate to. Particles such as electrons seem to us to be tiny spheres rather than slices of string because we are examining the particles from perpendicular dimensions in only three dimensions.

In my Theory of Stationary Space, our brane is not a three-dimensional space suspended in higher dimensional space. It is merely our instantaneous perception in four-dimensional space. We can only see in the three dimensions that we are used to. If we could see into the fourth dimension, we would just see the strings off into the distances of past and future like looking down a highway.

Our bodies, like all matter, consist of strings. At each point on the bundle of strings that is your body, your bodily and cranial processes produces a "particle" of consciousness. Once you have had your particle of consciousness at any particular point on the bundle of strings that is your body, you cannot have it again.

You must go on to the next point in the line and get the next particle of consciousness there.

However, we have a capacity called memory. This makes it seem to us as if time is a continuous process when, in fact, it is only our own creation. The past still exists but we have already had our particles of consciousness from those points. We could not see the world as we saw it then.

If my Theory of Stationary Space is correct then we really live in four dimensions of space, instead of three dimensions of space and one of time. Matter consists of long strings but the strings are aligned primarily in one direction. This dimension is what we perceive as time and the other three we see as space.

We can only see in the three spatial dimensions because our consciousness consists of a series of particles produced by the bodily and cranial functions of the bundles of strings composing our bodies. We cannot see in the fourth dimension, the one in which our bodily strings are aligned, but our series of consciousness particles cause us to perceive this fourth spatial dimension as time.

Incredibly, the secret to time travel is within our bodies and cranial processes because that is what defines time and determines our progress through it. Since it is our bodies and brains that actually create and define time, traveling in time would be somewhat like having an out of body experience. We cannot stand still in time at will because we can only get one particle of consciousness from each point on the strings composing our bodies.

POSTULATE 2—WE CANNOT MOVE AT WILL IN TIME SIMPLY BECAUSE IT IS THE DIRECTIONAL ORIENTATION OF THE STRINGS COMPOSING OUR BODIES THAT DEFINES TIME.

There is one thing that always brings back memories of my very early childhood. Our house, in the village of Lydbrook in Gloucestershire, was part of the way up the side of a hill. After I was in bed at night, I could see a vertical bar of light moving across the wall opposite the window. This was because there was a slight gap between the curtains and the headlights of a car moving on the road down below caused the vertical bar of light to move across the wall.

Forty years later, I was working as a telemarketer in America. It was Monday, March 15, 2004. I had gotten an appetite and looked at my watch to see how much longer it was to lunchtime. It was about 11 AM, I still had an hour to go. I had been reading quite a bit of physics and String Theory lately and I started to ponder that age-old question of what exactly is time. I looked out the window at the other buildings nearby.

What exactly, aside from motion, made one moment different from the next? Surely, there must be an answer. Since time was so primal to us, the answer was probably something relatively simple. Suddenly, that vertical bar of light seemed to shine on me. Of course, that was it. That is how our consciousness operates and it causes us to see the universe as we do. In a month or so, I had an entirely new theory of the universe worked out that explained both time and the speed of light and from there, the solutions to other mysteries of the universe just fell into place like a jigsaw puzzle. The thing that really gave away that time is really hidden space is the fact that it is affected by gravity just as space is.

Imagine traveling on a train. Except that your window to look outside is covered. They have placed two vertical pieces of cardboard over the window but there is a narrow slit between the pieces. You get to see a very narrow vertical slice of the scenery rushing by the train. At any given moment, the telephone wires running alongside the train look like a small ball bouncing up and down. In the distance are buildings on a street but through your narrow slit, you can only see one building at a time. Each building seems to "knock" the previous building out of the way. If this were all you could see of reality, you would never imagine that the previous buildings still existed.

One thing that is exceedingly difficult for most people is to visualize more dimensions than we are accustomed to. I wish to emphasize that you do not need to visualize four dimensions to understand this theory. A straight line is one dimension. It requires only one piece of information to locate a point on the line or to describe it's structure. A piece of paper is two dimensions. It requires two pieces of information to locate a point on the paper. A box contains three dimensions. It requires three pieces of information to locate a point in the box. In four dimensions, four pieces of information would be required to locate any point.

To travel backward in time under this theory, we would have to find a way to loop our body's strings backward in the fourth dimension but since we can only see and perceive in three dimensions. That will be difficult to do. And by doing this, we will still have to travel back at the same rate we travel forward. As for traveling forward in time, that is what we are doing anyway and if the bundles of strings making up our bodies are straight as it is, there would be no possible way to get there any faster. A better place to look for methods of time travel is within our own bodies and brains.

WHAT ABOUT ENTROPY?

One question soon arises. If time does not really exist, how can we explain entropy? This is a scientific precept that proves nature favors the simple over the complex and that a complex system is very unlikely to form from simplicity by itself but a complex system is very likely to break down into simplicity by itself.

One implication of this is that the so-called "arrow of time" can never be reversed. In other words, time can move in one direction only. If my Theory of Stationary Space is correct and time does not really exist, there must be a way around entropy and it's precept that time must move only in one direction and is most certainly not reversible.

Let's take a look at some examples of entropy.

Henry sits down at a table to have a cup of tea. He accidentally knocks the cup of tea off the table and it shatters on the floor. This is a perfectly plausible series of events. Now, if time was reversible or if it did not exist, then why do we not occasionally see shattered cups on the floor reassembling and jumping back up on the table? Obviously, so it seems, time is not reversible.

A bull just happens to stroll into a china shop to look around and, as bulls in china shops tend to do, leaves devastation in his wake. Next, a goat wanders into the shop and steps on the broken pieces but this time, the shattered china reassembles itself on the shelves. The first part of this story is somehow much more believable than the second. Clearly, the entropists say, time is not reversible.

To demonstrate entropy to his students, a science teacher fills an aquarium with water. In the aquarium, he places a bottle of dye so that the dye can escape from the top of the bottle into the water of the aquarium. Overnight, the dye mixes with the water of the aquarium. Day after day, the class examines the aquarium. Most of the dye never seems to go back into the bottle. It is clearly much easier for the dye to leave the bottle than to go back in. This can only mean that entropy makes time irreversible. For time to be reversible, the dye would have to go back into the bottle without any work being done to make it do so.

Now, let's take a closer look at the above examples of entropy. These examples all involve either living things or items made by living things. In fact, every example of entropy that I have ever read about seemed to involve either living things or items created by living things (such as the bottle). Living things do work to create complex systems out of simplicity. Have you ever read about or seen a meaningful example of entropy that did not involve either living things or items made by living things? I have not.

I believe that in my scenario in The Theory of Stationary Space, time could just as easily have ran in the opposite direction as it does. The only bundles of strings that would require any adjustment for this to occur are living things. Finally, if time is reversible or non-existent, there would be no requirement for the basic forces of nature, such as gravity, to reverse. It would be like walking down the street in the other direction. The only real difference would be that the universe would seem to be contracting rather than expanding, as it does now.

The classic examples of entropy involving teacups and china shops are somewhat silly. Entropy is meaningless outside the domain of living things and time, or at least our perception of time, could just as easily flowed in the other direction. Therefore I conclude that this theory is not hindered by entropy.

CHAPTER 3

▼

THE SPEED OF LIGHT

Aside from time, there is another great mystery that modern science seems unable to explain. This mystery is the speed of light. We can measure it very accurately. For our purposes, 186,282 miles per second or 300,000,000 meters per second is close enough. The great mystery is why is this the speed of light? Why does light, and any other electromagnetic radiation, travel at this speed in a vacuum and not at some other speed?

Einstein's Special Theory of Relativity emphasizes that the speed of light in the vacuum of space is absolutely invariable. Even time slows down for something traveling at close to the speed of light but the speed itself remains written in stone.

This makes it even more baffling, at least to me, why light travels at this particular speed and not some other. Why is the speed of light so absolutely fixed when even time is variable? If you are like most people, this is probably not the most pressing concern of your life, but have you ever stopped to wonder about it?

The reason that this is such a mystery is that there seems to be no physical reason whatsoever why light travels at this particular speed. Is it something to do with the nature of space itself? I searched in vain for some kind of connection between the speed of light and the universal gravitational constant. Is there some type of "speed bumps" in space that we have not detected that keeps light from traveling any faster?

I know that there is an electrical constant and a magnetic constant that gives the speed of light when multiplied together. However, I became convinced that this was not a good reason why light in a vacuum invariably traveled at this particular speed and that these two constants were the result of a division, or the "square root" of the speed of light.

A related issue is what cosmologists refer to as the "Horizon Problem". The universe is isotropic (essentially the same) in regions that a signal traveling at the speed of light could not possibly have reached from each other. How is this possible if light, and thus information, is limited to a certain speed?

THE SPEED OF LIGHT DOES NOT EXIST

The truth is that there is no speed of light. Just like time does not really exist, neither does the speed of light. This naturally makes sense if time does not really exist since such a speed as that of light is a function of time. The two great holes in modern physics, our inability to explain either just what time is or why light travels at this particular speed, is because neither really exists.

But what on earth am I talking about? We can measure the speed of light with great precision and, unless it travels through air, water or glass, it always comes out the same. Even if we cannot yet find some physical reason for this, surely it must exist.

We saw earlier how our view of the universe is the three spatial dimensions that we see while our series of particles of consciousness produced by our bodily and cranial processes moves continuously along the bundles of strings making up our bodies. These strings, which when seen in our familiar three dimensions are the quarks, electrons and, other particles making up all matter, are aligned along the fourth dimension.

Thus, we cannot see along this dimension as we can in the other three. If we could, we would be seeing backwards or forwards in time. We cannot move in the past because we have already had our particles of consciousness there and we cannot move in the future because we have not yet had our particles of consciousness there.

POSTULATE 3—WHAT WE PERCEIVE AS THE SPEED OF LIGHT IS ACTUALLY THE RATE AT WHICH OUR CONSCIOUSNESS MOVES ALONG THE BUNDLE OF STRINGS COMPOSING OUR BODIES AND BRAINS.

The speed of light does not exist. We can find no physical evidence as to why light moves at this particular speed through space and not some other speed. When we measure the speed of light, what we are really measuring is the rate at which our series of particles of consciousness moves along the bundles of strings making up our bodies.

This speed, in the fourth dimension, causes us to perceive that this is the maximum speed of any motion in the three dimensions that we see. If we moved through consciousness at some other speed, that speed is what the speed of light would seem to be. Einstein was right that there was a close relationship between time and the speed of light but he did not know why because he was thinking in only three spatial dimensions and was not thinking about the real nature of consciousness.

We can measure the speed of light with great accuracy. But it is not really what we are measuring. When driving down the highway, trees appear to be moving apart or coming together due to our relative motion. If we assumed the car to be stationary and the scenery to be in motion, we could measure by triangulation and time exactly how fast the trees were "moving". But no matter what, we would never get a figure faster than the car was moving. So it is with what we perceive as the speed of light.

CHAPTER 4

▼

CONSCIOUSNESS

The bundles of strings composing our bodies, which in three dimensions we see as quarks, electrons and, other such fundamental particles, produce particles of consciousness. Time flows, but it flows in the same way that the sun will rise and set, it is only something that we perceive. Our flow in the fourth spatial dimension, the one that we perceive as time, is the result of how the "motion" of the particles in our bodies and brains results in a "particle" of consciousness. This particle is followed by another and another and so on.

The best analogy of consciousness that I can think of is the burning of a fuse. Just as the fuse produces a series of sparks as it burns along, our perception of time is our series of particles of consciousness. We must move along the fourth (time) dimension to have consciousness because it requires motion of the atoms in our bodies to maintain consciousness. If we could attain and maintain consciousness with less motion, time would appear to go slower. If we could have continuous consciousness from only one point on our bundles of strings, there would be no apparent motion in the universe.

It is the internal processes in a car that determines how fast it goes and thus how fast the scenery "rushes by". So it is with our consciousness moving along the bundles of strings making up our bodies and brains. We are "traveling" at what we perceive as the speed of light but it is only our consciousness that is traveling and it is not traveling in the other three dimensions that we perceive as space.

The particles in the atoms of our bodies are, of course, not really in motion but are not perfectly straight and so are in relative motion as our consciousness rushes by. This relative motion is what gives us our bodily and cranial processes. Just as a book cannot really "tell" you a story because it is an inanimate object with static and lifeless letters, your consciousness moving over the pages of the book gets the same result as if the book "told" you the story.

The motion of your consciousness at 186,282 miles per second over the static strings composing your body and brain creates the apparent motion of the strings that creates the next particle of consciousness. Of course, since you can experience only three of the four dimensions as space, you perceive yourself as having a three dimensional body moving in one dimension of time.

WHY THE "SPEED OF LIGHT" SEEMS TO US TO BE THE MAXIMUM POSSIBLE VELOCITY

Probably a better example than a car speeding down the highway is a ship moving past land. Everything on the land appears to be in motion while the ship seems to be standing still. If we used surveying equipment along with a timer to measure how fast things were "moving" on the land, we would never get a figure that was faster than the ship was moving (assuming, of course that everything on the visible land was actually at rest).

POSTULATE 4—OUR CONSCIOUSNESS IS PROGRESSING ALONG THE STRINGS OF OUR BODIES AT WHAT WE PERCEIVE AS THE SPEED OF LIGHT. EVERYTHING ELSE IS AT REST. THIS IS WHY WE PERCEIVE THAT NOTHING IS ABLE TO MOVE FASTER THAN THE SPEED OF LIGHT.

This is exactly how our consciousness causes us to perceive the speed of light, the maximum speed limit of everything in motion, but why we can find no physical evidence of just why this is the speed of light. It is because it does not exist. It is a trick that our consciousness plays on us because our consciousness is progressing at this speed along the strings composing our bodies and we can only see in three of the four dimensions. Our consciousness progresses at this particular speed because it is sufficient to provide consciousness from the static strings composing our bodies and brains.

DISTANCES ALONG THE FOURTH DIMENSION

This means that time, as we perceive it, is actually distance along the strings aligned in the fourth dimension. Since it requires great complexity of stationary strings to provide us with even a moment of consciousness, time is equivalent to great distances along the strings.

This should not take us by surprise since the distances we require along the strings of the fourth dimension to support the history of the world and the universe as we know it is no greater than the distances that we can observe across in the three spatial dimensions. Although if I had come up with this theory in 1915, it would have seemed like nonsense due to the size the universe was thought to be at that time. There would not have appeared to be enough room to accomodate even human history if every second of time required 186,000 miles.

Obviously, a second of time corresponds to 186,282 miles along the strings. Incidentally, this is 10,317,157 times as fast as a car traveling at 65 miles per hour. This means that every minute, your consciousness progresses 11,176,920 miles along the bundle of strings composing your body and brain. An hour would equal 670,615,200 miles. A day is equivalent to 16,094,764,800 miles. A year would be 5,880,000,000,000 miles (also known as a light year). A lifespan of eighty years would be 470,000,000,000,000 miles. Two thousand years would equal 11,760,000,000,000,000 miles of distance. This is really life in the fast lane. Roller coasters will seem boring from now on.

HOW WE EXPERIENCE FOUR DIMENSIONS

The particles that we are familiar with, quarks, electrons and so on, are one-dimensional strings existing in the four-dimensional space and aligned in one direction, the dimension we perceive as time. When quantum theorists examine particles in our three-dimensional space, they see a cross-section or one point at a time of the string, which they perceive as an infinitesimal point particle.

Our familiar three dimensions are a one-dimensional slit of four-dimensional space. Consider a piece of paper in a filing cabinet. The space in the cabinet is three-dimensional and the piece of paper is two-dimensional. If we could look at the paper edge-on and could only see inside the piece of paper, we would only be able to see in two dimensions even though the filing cabinet contains three dimensions of space. Since space has four dimensions of which we see a one-dimensional slit, that means we live in three dimensions.

This is how we see the four-dimensional universe we live in. The three-dimensional slit of it that we can see corresponds to the piece of paper in the filing cabinet. What this means, of course, is that our universe is larger in terms of space than we can possibly comprehend. If you have a struggle to grasp the size of the observable universe now, about 15 billion light years across, try to imagine adding another dimension to it. This means that the universe is almost infinitely larger than we perceive at a given moment because it includes all of the past and future.

POSTULATE 5—WE CAN SEE ONLY THAT PORTION OF THE UNIVERSE THAT IS PERPENDICULAR TO THE POINT ON THE STRINGS MAKING UP OUR BODIES WHERE OUR CURRENT PARTICLE OF CONSCIOUSNESS IS LOCATED. SINCE OUR UNIVERSE HAS FOUR SPATIAL DIMENSIONS, THAT MEANS WE SEE IN THREE DIMENSIONS. THE INFINITESIMAL SLICE OF THE FOURTH DIMENSION WE SEE IS WHAT WE CALL THE PRESENT.

We see not reality but that slice of reality that happens to be perpendicular to our present point of consciousness. If we could see in the other dimension, we would be looking backward or forward in what we know as time. We see the fundamental particles as points because they are one-dimensional strings aligned along the dimension we perceive as time and we are seeing them only in a perpendicular dimension.

Where a string begins and ends is when a particle apparently begins to or ceases to exist. But it only seems that way to us because we are seeing only a moving one-dimensional slice of a four-dimensional universe. The motion is at what we call the speed of light and is caused by the series of particles of consciousness that make up time to us.

All that we see and interact with during our lives is merely a one-dimensional slice of a four-dimensional universe. This gives us our familiar three-dimensional universe along with the perplexing mysteries of time and the speed of light. Our consciousness is somewhat like an electron moving along a wire but only able to see perpendicular to the wire.

LIVES IN TIME: RELATIVITY OF THE PRESENT

Whenever you interact with another person or persons, it is clear that you are both having your moment of consciousness at the same time. This would make it seem that there must be a kind of "consciousness frontier" or "consciousness wave" in which the sparks of consciousness of all persons alive at that time are occurring along a "frontier" that is perpendicular to the dimension we perceive as time. Otherwise, few of the people in the world would be having their moments of consciousness perpendicular to yours so that you could interact without one of you being "alive" while the other was "dead".

It is possible that such a "Frontier Model" as I will call it is valid. However, remember that time does not really exist. Our bodies and minds are composed of bundles of strings and time is really distance along the fourth dimension. I prefer what I call the "Different Points Model". Whenever you have a live interaction with another person, you are interacting with that person the way they are perpendicular to the point where you are having your present moment of consciousness.

I believe that the other person may not necessarily be at that moment of consciousness themselves, you are seeing and interacting with them as they were in their past or will be in their future. Remember that you are alive all your life, all the trillions of miles of it, but only conscious from your own perspective at one point at a time.

There is no way you can know where a person's present moment of consciousness is when you talk or otherwise interact with them, it may be in the direction of either the past or future from your present point of view. All you can be sure of is that you are at your present moment of consciousness when the interaction takes place.

The other person registers the interaction when they are experiencing their moment of consciousness at a point perpendicular to where you are having your moment of consciousness when the interaction takes place. That point, however, may be in the person's past or future depending on where they are having their moment of consciousness when you are having yours during the interaction. A person could actually be long dead from the point of view of their own consciousness when you talk with them.

If this seems bizarre, remember that time does not really exist except in our consciousness. This means that your birth and death (or rapture) happened at essentially the same time and your entire life, comprising all of your moments of consciousness, always exists. Another person can interact with you at any point during your life and both of you will perceive the interaction as the present although the other person may be long past the interaction in terms of moment of consciousness or may not have arrived there yet.

Einstein pointed out that time is relative but this is not what he was talking about and it is, in fact, more relative than he ever imagined. We are all passing through the same world and universe but the actual date is different to each one of us. We could call this "Relativity of the Present".

If two people were born on the same day, the bundles of strings composing their bodies will be intact roughly parallel to each other but that does not mean that the movement of consciousness down their bundles will take place concurrently. I may chat with you on June 15, 2004. Of course, it will be the same date for you when the chat takes place. But my moment of consciousness may be in the future relative to yours, to you it may still be 1979.

In that case, you will not have the chat with me until you get to June 15, 2004, even though both of us will perceive it at the present when the chat takes place. At that point, I will be in 2029 and will recall the chat with you of June 15, 2004. Our chat will exist essentially forever long after we are both past it. Remember that there is really no such thing as time except to our consciousness and the past, present and, future exists all at once, separated only by spatial distance.

Your moment of consciousness may be ahead of mine in the movement along the fourth dimension and you may remember cutting me off in traffic. But I will not experience you cutting me off in traffic until my consciousness gets to that point. The reason for this is simply that time does not really exist and thus your past, present and, future always exists.

We are "alive" all our lives but we are only conscious of it one point at a time or, what we perceive as time. That point is the present to us but if someone should interact with us whose consciousness point is not perpendicular to ours, they will nevertheless interact with our consciousness at the point on the bundle of strings composing our body. We were or will be conscious at that point, even though we are not at that moment of consciousness now, so each person perceives the other as having their moments of consciousness concurrently.

THE NATURE OF THE OBSERVER

I believe that it is necessary to make the observer a part of any theory about the nature of the universe. We already know about the Anthropic Principle that limits acceptable cosmological theories to those taking human history into account. Obviously, our models of the universe must make it possible for life to emerge or we would not be here to ponder the universe.

Quantum Physics implicates the observer into the physics and in this respect is ahead of cosmology. In Quantum Mechanics, it has long been known that the very act of observing can change the outcome of an observation. According to the Copenhagen Interpretation the wave function of a particle is a complete description of that particle, meaning that the best we can really know about particles is probabilities. The particle is referred to as "superposed". It is the observer who actually turns it into a particle or a wave.

There was a well-known thought experiment concerning a cat in a box with a cyanide capsule dreamed up by a physicist named Erwin Schrodinger. If the cat has bitten into the capsule, it would naturally be dead. If it had not bitten the capsule, it would be alive. Schrodinger's cat in the famous box paradox is believed to be neither dead nor alive until the box is actually opened. It is "superposed" until it is actually observed. In the so-called Post-Everett interpretation of Quantum Physics, "decoherence" means that an event is at least potentially observable.

In Quantum Physics, observability is understood to be a fundamental factor in experimentation. The observer may actually determine the result of an experiment or observation merely by observing it.

If Quantum Physics includes superposing, decoherence, as well as the well-known Heisenberg Uncertainty Principle, then why should not the observer, as well as the nature of the observations, have a significant role in cosmology also? Quantum physicists know better than to trust their senses too much. I believe that cosmologists need to accept this too. When an astronomer looks into space through a telescope, his eyes and brain become part of the instrument also. His interpretation of what he sees is a vital part of the observation.

We must keep in mind that our senses were not primarily designed for cosmology. The universe may not really be the way our senses tell us it is. The way the universe appears to us is the result not only of what it is but of what we are.

POSTULATE 6—WE KNOW THAT THE OBSERVER PLAYS AN IMPORTANT ROLE IN QUANTUM PHYSICS. THE OBSERVER ALSO PLAYS A VITAL ROLE IN COSMOLOGY. IT IS IMPOSSIBLE TO UNDERSTAND THE UNIVERSE WITHOUT UNDERSTANDING THE ROLE THAT OUR OBSERVATION PLAYS IN IT.

At a glance, the earth looks to be flat. The sun appears to go around the earth. Gravity seems to be a force. The sky appears blue. The world around us is full of color. Yet, we know that all of this is illusion.

Color, by the way, does not really exist. Certain wavelengths of light are interpreted by our eyes and brains as red, others as green. But the color does not actually exist until the light reaches our eyes. The sky is not really blue, it is merely the average color of the mixture of light refracted to us by the atmosphere.

The essence of The Theory of Stationary Space is that time and the speed of light are merely the products of our interpretation of the world around us in much the same way. The two are illusions.

POSTULATE 7—THE FURTHER WE GET FROM OUR EVERYDAY WORLD THAT WE WERE DESIGNED FOR, EITHER TO THE REALM OF THE VERY LARGE OR THE REALM OF THE VERY SMALL, THE LESS WE CAN RELY ON OUR OWN OBSERVATIONS AT FACE VALUE REGARDLESS OF THE QUALITY OF OUR INSTRUMENTS.

CHAPTER 5

▼

MOTION

The strings making up the matter in the universe are set in four-dimensional space primarily in one direction. This direction we perceive as time. The other three dimensions we perceive as space. We cannot move at will in the time direction because the strings making up matter are parallel to it and we, of course, are made of matter.

What we perceive as motion is nothing more than variations in the straightness of the strings when our consciousness comes rushing through at what we perceive as the speed of light. Since our consciousness is moving along the strings at such high speed, the variations in the straightness of the strings, which we perceive as motion, would be almost imperceptible if seen from outside in four dimensions, but adds up gradually over hundreds of thousands of miles along the strings.

Time does not really exist but is merely our perception. Motion, which is a function of time, does not really exist either. Motion is merely static bumps in and variations in the straightness of the strings and bundles of strings. Just as time is only apparent time and the speed of light is just the apparent speed of light, motion is just apparent motion because our consciousness is rushing past strings that are not perfectly straight at very high speeds and we see only a one-dimensional slice of the fourth dimension of four-dimensional space.

POSTULATE 8—IF TIME DOES NOT REALLY EXIST THEN MOTION, WHICH IS A FUNCTION OF TIME, CANNOT REALLY EXIST EITHER. WHAT WE PERCEIVE AS MOTION IS VARIATIONS IN THE STRAIGHTNESS OF STRINGS OR GROUPS OF STRINGS AS OUR CONSCIOUSNESS PASSES BY.

FOUR-DIMENSIONAL EXPLANATIONS OF EVENTS WE PERCEIVE IN OUR THREE DIMENSIONS

Orbits such as a moon around a planet or an electron around an atom, is really strings wrapped like a cable. As our consciousness rushes past and we only see perpendicular to the direction our consciousness is traveling, a small string or bundle of strings wrapped around a larger bundle appears to us as if it is orbiting the larger bundle. We could easily calculate the wrap length using what we know as the speed of light.

What we perceive as an explosion is actually the strings in a bundle of strings coming apart at certain angles. The sharper the angles, the more powerful of an explosion we would perceive. Explosions only appear to us to happen because our consciousness is moving at such a rate. An explosion is just a bundle of strings diverging apart. It would not appear so explosive if our consciousness was progressing more slowly.

Consider a ball falling from a height of sixteen feet to the earth's surface. It would take exactly one second to make the fall. If we could look at this in four dimensions, we would see two parallel bundles of strings. There would be a very large bundle, the earth, and a small bundle, the ball. At the start of the fall, the ball bundle would be sixteen feet away from the earth bundle. If we moved in the direction that our consciousness is traveling along the strings, the ball bundle would very, very gradually get closer to the earth bundle. After we had gone 186,282 miles from the point where the "fall" began, the ball bundle of strings would finally meet the earth bundle. How incredibly different this would appear from the sight in three dimensions of a ball dropping to the ground from a height of sixteen feet in one second. This is because we are seeing in only three of the four dimensions and our consciousness is moving at such a velocity.

Suppose we could look at Niagara Falls in four dimensions. We would have a large bundle of strings, which is the earth. There would be an escarpment on the bundle 175 feet high (for the Horseshoe Falls). There would be other strings that would be at the height of the escarpment but would gradually diagonally get closer and closer to the bottom of the escarpment. These strings (the water over the falls), would each take more than half a million miles along the escarpment to reach the bottom of the escarpment after having started from the top. When you watch the water drop, you are seeing only a moving (at 186,282 miles per second) three-dimensional slice of the four dimensions. Remember, parallel strings we perceive as stillness while diagonal strings we perceive as being in motion but the truth is that both are equally still.

Keep in mind that in four dimensions, instead of our three, there are only four dimensions of space. There is really no such thing as time except for the movement of our consciousness along the bundles of strings making up our bodies. Our consciousness is so intricate that it requires 186,282 miles of strings to provide us with what we perceive as one second of consciousness. This gives us the illusion of both time and motion in an essentially still and timeless universe and causes us to perceive the speed of our consciousness as the speed of light, when in fact there is no such thing.

In my Theory of Stationary Space, the only new movement in the universe since the universe was laid down is that caused by living things. In fact, life can be defined as that which creates new arrangements of strings. Now that we know this, it becomes clear why the only meaningful examples of entropy involve living things and items made by living things, such as bottles. It is because the consciousness in living things is moving in one direction only.

POSTULATE 9—THE REASON THAT MEANINGFUL EXAMPLES OF ENTROPY ARE FOUND ONLY IN LIVING THINGS AND ITEMS MADE BY LIVING THINGS IS THAT LIVING THINGS COMPRISE THE ONLY NEW MOTION IN THE UNIVERSE AND THE CONSCIOUSNESS IN LIVING THINGS IS MOVING IN ONE DIRECTION ONLY.

When two bundles of strings meet, we perceive it as a collision. However, this is only when our consciousness passes that point. The meeting of the two bundles (the collision) is actually timeless. It existed since the universe was laid down and will continue to exist indefinitely after our consciousness is long past it.

Strings (Seen as fundamental particles in three dimensions) and bundles of strings (Seen as objects in three dimensions) affect and interact with each other but it is a stationary and permanent interaction, not a collision as we perceive. The bounce of a ball is timeless. It always exists. It is just that our consciousness moves on past it.

Our memory tricks us into thinking that time really exists in the universe out-side our bodies when it does not. At the same time, our memory also gives us an idea of what the four-dimensional universe really looks like. It all exists at once. Not only has everything already happened but, it all happened at once. We are just experiencing it in the way one reads a book as our consciousness progresses along the strings composing our bodies in the fourth spatial dimension that we perceive as time.

A bouncing ball, as seen in three dimensions, is not really a ball. It is a cylinder the same diameter as the ball but it is probably trillions of miles long and it extremely gradually meets and moves away from the large bundle of strings that is the earth in four dimensions. An object, such as the ball, that lasts ten years, as we perceive it, is almost fifty-nine trillion miles long but we see it as one sphere at a time in our three dimensions. This is what we perceive as time.

Suppose we, in our three dimensions, aim a radar pulse at the moon so that it will bounce back to us. What we are actually doing in four dimensions is aiming the pulse so that it will bounce off the moon and be where our consciousness will be when we get to that point. When you shine a light on something and see the reflected light, you are actually shining the light toward the future in four dimen-sions. Even though it will appear in three dimensions to be going straight to the object and reflecting straight back to you.

Light travels in all four dimensions even though we are confined to three dimensions. Light is permanent ripples in space and travels essentially instanta-neously. It is our consciousness that travels at 186,282 miles per second along the bundle of strings composing our bodies.

What about Newton's Laws? The reason that a body of matter in motion remains in motion and a body at rest remains at rest until acted upon by an out-side force can be easily accounted for by the straightness of the strings composing a body. If your body is composed of a straight line of strings and another line is set at a certain angle to your body's strings, the body of matter represented by the other line will appear to you to be in constant motion. Likewise, another body of matter represented by a bundle of strings running parallel to the bundle of strings composing your body will seem to you to be a body of matter at constant rest.

POSTULATE 10—NEWTON'S LAWS OF MOTION CAN BE EXPLAINED BY THE STRAIGHTNESS OF THE BUNDLES OF STRINGS COMPOSING BODIES OF MATTER. THE LAW OF INERTIA BECAUSE FROM THE POINT OF VIEW OF A STRAIGHT LINE, ANOTHER STRAIGHT LINE WILL SEEM TO HAVE A CONSTANT

RATE OF MOTION (OR STILLNESS) WHEN VIEWED FROM SUCCES-
SIVE POINTS ON THE ORIGINAL LINE TO CONGRUENT POINTS
ON THE SECOND LINE. NEWTON'S LAW CONCERNING OPPOSITE
AND EQUAL REACTIONS BECAUSE THE AVERAGE DIRECTION OF
THE STRINGS COMPOSING MATTER ALIGNED IN SPACE IS
ALWAYS CONSERVED.

FOUR-DIMENSIONAL EXPLANATION OF E=MC SQUARED

In The Theory of Stationary Space, any kind of "motion" actually converts matter into energy. This is just as Einstein stated in his famous formula E=MC squared. In our theory, what we perceive as time is actually the "matter dimension" and the other three are "energy dimensions". Motion, and thus energy, is any twisting or bending in the strings. We perceive it as motion and energy as our consciousness goes rushing past it. There is no actual energy or motion, only apparent motion because our consciousness is moving.

Matter and energy is interchangeable because a twisting of a string causes "motion". But, it decreases the overall length of the string in the matter (time) dimension. That is why Einstein pointed out that matter and energy are interchangeable, although he did not know why.

If you stretch out a length of string on the floor, it will reach further than if it had slight bends in it and was not perfectly straight. If it is not perfectly straight, representing matter, the string sacrifices some length to go off in a perpendicular dimension, representing energy.

The speed of light enters into Einstein's famous formula, E=MC squared because that is the rate our consciousness is progressing along the strings composing our bodies. C in the formula stands for "constant' or the speed of light, which is always constant in relativity. Multiplied by itself, or squared, is the same speed of consciousness (light) in the dimensions perpendicular to the dimension we perceive as time.

The speed of light is in the formula twice and is multiplied by itself, or squared, because that is the speed our consciousness is progressing along the fourth, or matter, dimension and it is thus the speed at which an object would appear to us to be traveling, or it's strings would appear to come apart, for us to perceive it as totally converted from mass into energy. This is because the strings composing the object would be bent at right angles to the strings composing our bodies, so it would appear to us to have been entirely converted into energy, or motion at the speed of light, from stationary matter.

So, we get the speed of light twice in the formula, once for the velocity of our consciousness and once for the apparent velocity that matter would appear to us to be traveling if it were completely converted into "energy". This is simply because the strings of that matter would be bent at right angles to our strings and would thus exist only in the perpendicular, or energy, dimensions from that point on if the string appeared to us to be traveling at that speed. The two speeds of light are multiplied together, or squared to get the function in the formula.

Einstein stated that matter and energy are interchangeable but now we can see that this is simply because strings can be aligned along either the one dimension we perceive as matter or time and the other three we perceive as energy or motion. It would be more correct to say that time and distance are interchangeable because time actually is distance.

Completely converting mass into energy would be tantamount to placing the strings of the mass perpendicular to the dimension we perceive as time. The mass would appear to us to have vanished. Just as if all strings were perfectly straight, it would be perceived by us to be motionless and at a temperature of absolute zero.

FOUR-DIMENSIONAL EXPLANATION OF RELATIVITY

When a string or bundle of strings is parallel to the bundle of strings that is your body, it will appear in our three dimensions as an object at rest. When the bundle of strings is not parallel to our bodies, it appears to us as an object in motion. The further from parallel it is to us, the faster it appears to be traveling. If the bundle of strings bent enough so that it was perpendicular to our bodily strings, it would appear to us to be traveling at what we call the speed of light or to have been totally converted from matter into energy. But this is only because it is the velocity at which our consciousness is traveling.

The object would seem to us to have infinite mass as it appeared to travel at the speed of light simply because all of the mass of the bundle of strings would seem to be concentrated at one point and as our consciousness passed it, would seem to no longer exist. This is another mystery solved, why an object would have infinite mass if it traveled at the "speed of light", according to relativity.

Einstein pointed this out and theorized that it would naturally be impossible for the object to travel any faster because it would require an infinite force to do so. The truth is that the object (bundle of strings) would seem to us to have infinite mass only because it would be perpendicular to us as our consciousness passed it and thus it's millions or trillions of miles of mass would seem to us to be concentrated at one point.

POSTULATE 11—STRINGS PARALLEL TO THE STRINGS COMPOSING OUR BODIES WILL BE PERCEIVED BY US AS BEING AT REST. STRINGS PERPENDICULAR TO THE STRINGS COMPOSING OUR BODIES WILL BE PERCEIVED BY US TO BE TRAVELING AT THE SPEED OF LIGHT.

THE APPARENT VELOCITY IN THREE DIMENSIONS IS GIVEN BY THE SINE OF THE ANGLE BETWEEN THE TWO MULTIPLIED BY THE APPARENT SPEED OF LIGHT. THE TIME DILATION IN THREE DIMENSIONS IS GIVEN BY THE COSINE OF THE ANGLE BETWEEN THE TWO.

Einstein pointed out that nothing is able to travel faster than the speed of light. He was correct but was only thinking in our familiar three dimensions. The truth is that the objects are not actually moving at all. It is our consciousness that is moving along the strings composing our bodies at what we perceive to be the speed of light. When another bundle of strings is bent at a right angle to our strings, we perceive that it is traveling away from us at the maximum possible velocity, which we perceive as the speed of light. In fact we will no longer be able to see it. But it is because of the angle at which it is bent, not because of the speed at which it is traveling. This "speed" is but an illusion caused by the motion of our consciousness.

The reason that we perceive nothing as capable of traveling faster than light is simply that there can be no angle away from us greater than a right angle. A bundle of strings at a right angle to the strings of "stationary" objects would appear to us to be moving at the same speed that our consciousness is really moving. We perceive ourselves as being at rest but even when we are sleeping, our consciousness is moving along the strings composing our bodies and brains at what we perceive as the speed of light.

The reason that relativity does not set in until close to the speed of light is the relative insignificance in the distortion of the strings at lower speeds. We perceive that an object traveling at the speed of light would have infinite mass. But this is just another illusion of our three dimensions. The bundle of strings turns at a perpendicular angle away from the direction that our strings are aligned so we perceive that all of it's mass is concentrated at one point.

The slowing down of time at relativistic speeds, as pointed out in Einstein's Special Theory of Relativity can be explained in terms of simple trigonometric functions. Consider a radius, an x-axis and a y-axis in standard trigonometry. Suppose the bundle of strings making up our bodies is aligned with the x-axis. Now suppose that another bundle of strings making up an object such as a meteor is parallel to our strings. However, our consciousness reaches a point on our strings where the meteor bundle bends at a certain angle. In our three-dimensional viewpoint, we would perceive this as the meteor traveling at a certain velocity. The velocity would depend on the angle of the bundle of strings relative to the bundle making up our body.

The sine of the angle of the meteor bundle of strings would give it's velocity. The sine of an angle starts from zero at the x-axis and goes to one at the y-axis, or 90 degrees. The velocity to us would appear as the sine of the angle times what we perceive as the speed of light, in other words the speed of our consciousness.

Likewise, the cosine of the angle would give the time dilation. The cosine of an angle is the opposite of the sine. It starts at one at the x-axis and goes to zero at the y-axis, or 90 degrees. At the x-axis, parallel with our strings, time dilation would be zero. Time would seem to run normally at the x-axis. As the meteor's velocity increases, time seems to pass slower and slower until at the y-axis, perpendicular to the x-axis and with a cosine of zero, there is no sensation of time at all.

This is why, in our three-dimensional universe, we perceive that something moving at the speed of light seems to move between any two points instantaneously from it's own point of view. Time does not exist in perpendicular dimensions, even in perception. Someone traveling in a dimension perpendicular to our fourth (time) dimension would not be making any progress along the time dimension and thus, time would seem to stand still.

The reason for this, of course, is that the axis in which the strings of matter are aligned in the four-dimensional universe represents time (as we perceive it) or matter. The other three dimensions represent motion and space. But in the four-dimensional view, nothing is really moving at all.

The greatest angle that a string or bundle of strings can point away from us is perpendicular or ninety degrees. That would cause us to perceive that it is moving at the speed of light. For something to travel backwards in time, it would have to be pointed away from the dimension we perceive as time at greater than ninety degrees, which is naturally impossible. If a bundle of strings were pointed away from us at an angle greater than a right angle, it would appear in our past as a second object meeting and colliding with the first object and then ceasing to exist.

The reason that nothing can ever travel faster than the speed of light is simply that nothing is really moving at all and our consciousness is moving along the strings making up our bodies at 186,282 miles per second and so we perceive this as being the speed of light while we imagine ourselves to be at rest. If objects in the four-dimensional universe were actually moving, we would perceive objects as traveling faster than the speed of light from our three-dimensional viewpoint.

OUR PERCEPTION OF THE UNIVERSE

The way we perceive the universe is due to three great factors. First of all, there are four dimensions of space but we can only see in three of them. The fourth we perceive as time but it is something that has no real physical existence outside of our consciousness.

The second factor is the speed at which our consciousness moves down the strings making up our bodies. If we moved slower, the universe would seem to be a very different place, more slow-paced. If we moved faster, the universe would appear to us as more energetic and violent. But the truth is that the universe is not in motion at all, it is our consciousness that is in motion. The way we see the universe is due not only to what it is but to what we are.

The third factor is the straightness of the strings and the conservation of the average direction in which the strings are aligned in space. This explains Newton's Laws. A string or bundle of strings can move out of parallel alignment but it always causes another bundle of strings to move out of alignment in the opposite direction. This conserves the average directional alignment. Let's call it the "mean spatial alignment".

Suppose you visited friends or relatives in Europe. However, when they took you around to show you their village, you were taken to a bullet train that whisked you through the village in three seconds. All you really saw was a blur. It was only the objects further away that you saw with clarity.

This is about the way we are seeing our universe, except that we do not realize it because we only perceive it in three dimensions. Just as a rider on a fast train would only see objects further away with clarity, so with us it is the distant, almost unchanging stars that are our most realistic picture of the universe.

The universe is a very placid, still and, gentle place. It appears to us as violent and energetic as it does because our consciousness is rushing through it at such great velocity but we perceive ourselves as being at rest. If only we could look at the universe in four dimensions, it would appear utterly different than what we see in three dimensions. It would look as different as if we looked at a two-dimensional slice of our familiar three dimensions.

We would see four dimensions with strings rather than particles. The strings would be aligned almost perfectly in one of the four dimensions. Strings would very gradually, almost imperceptibly, move in relation to each other only if you looked thousands, or maybe millions or even billions of miles down the strings. What we in three dimensions perceive as violent collisions between objects would appear as very gradual meetings of parallel bundles of strings in four dimensions.

When you watch a moving object in our familiar three dimensions, you are really watching successive positions of a bundle of strings out of parallel to your body's bundle of strings through a one-dimensional slit of the fourth dimension. This is the only plausible way to explain time and it also explains why the speed of light is what it is and why C squared or, the speed of light squared, is Einstein's function for the conversion of mass and energy.

CHAPTER 6

▼

FORCES AND QUANTUM PHYSICS

GRAVITY

When we look in four dimensions, gravity appears as a deep well in the surrounding space running parallel to and surrounding a bundle of strings. Electromagnetic waves are much smaller and finer ripples perpendicular to bundles of strings. This explains why gravity bends light as well as why most objects are opaque, unless the ripples can move between or around the strings of an object. If gravity did not bend light, it would appear to go around massive objects. It would pass by above the gravity well. This is not a new idea, Einstein's Special Theory of Relativity described this, although not in four dimensions.

Since gravity is an indentation in space or a "trench" with the bundle of strings at the bottom, we can see that the so-called escape velocity of a massive object depends on the angle from the surface of the object in a straight line necessary to get to the level of space above the gravity well. We have already seen how the sine of the angle multiplied by what we perceive as the speed of light will give us the object's escape velocity.

Keep in mind that the bottom of the gravity trench (usually referred to as a "gravity well" in our usual three spatial dimensions) is drawn to the object's center of gravity and not to the physical bottom of the object. A larger body of equal mass will require a lower escape velocity because it's larger radius will provide a "head start".

What physicists call a "black hole" is simply a very massive bundle of strings that causes it's gravity trench to have sides steep enough so that the ripples we perceive as light cannot escape it's gravity in a straight line. Another possible explanation for black holes is that the bundle of strings are so massive that the fabric of space is stretched taut enough to prevent formation of the ripples we perceive as light, although I favor the first explanation. This would require reevaluation of the way we currently view black holes.

Of course, we experience very little gravity in the fourth dimension (the dimension of matter or time) due to the canceling out of equal gravitational pulls from opposite directions. This cancellation helps to conceal the fact from us that the fourth dimension is actually space, rather than time.

The only gravity in the fourth dimension is due to unequal pulls from past and future due to bent bundles of strings, what we perceive as motion and this must be very insignificant in comparison to gravitational pulls from the perpendicular dimensions. Strings normally align along the fourth dimension and gravity tends to reinforce this. The fabric of space is semi-taut. It is this tautness that makes the universe as it is and gives us the physical constants that govern the universe. Now, we can actually measure this tautness with the information contained in this theory.

ELECTROMAGNETIC RADIATION

Now that we understand motion to not really exist, it becomes clear that electromagnetic radiation can only be stationary ripples in the fabric of space caused by the slight changes of direction of strings. Even the sun on a hot summer day is only hot and bright because we are rushing through the stationary ripples in it. Space is a truly incredible fabric capable of holding very deep gravitational warps as well as many, many different ripples that we know as electromagnetic waves, going in all different directions, all at the same time.

Of course, the stars do not really shine. It is an illusion of our motion. We are rushing at 186,282 miles per second through the ripples in the fabric of space that we perceive as light and that makes it look as if the stars are shining. Light and other electromagnetic radiation, as we understand it is caused by motion, whether the motion of electrons up and down an antenna or the dropping of an electron in an atom from a higher to a lower orbital causing the release of what we perceive as a pulse of energy.

Sound is permanent and stationary waves in matter caused by the meetings of bundles of strings. These indentations sound loud to us only because our consciousness is going through them at such high velocity. Sound seems to us to be nowhere near as fast as light simply because sound is ripples in matter, which means the sound is really going in almost the same direction our consciousness is going and meets us very gradually, while light is ripples coming to us at right angles and we perceive it as traveling at the speed our consciousness is actually traveling. Unlike light, the apparent speed of sound is determined by the nature of the host matter.

The bending or changing of direction of a string or group of strings causes ripples in perpendicular directions through space. It is these pervasive ripples that we perceive as electromagnetic radiation. The ripples are actually static, just like the strings composing matter. It is our consciousness that is moving through the ripples at 186,282 miles per second. The result is that we see the ripples as moving that fast while we perceive ourselves as stationary, just as someone on a large ship near land will perceive the land to be in motion while the ship is stationary.

An inferno is just a stationary jumble of strings. It would not appear to be so infernal if our consciousness was moving more slowly. Of course, there is no such thing as temperature. Heat is stationary waves and tangles in strings that our consciousness rushes by and perceives as rapidly moving particles and the resulting ripples in the fabric of space that we perceive as light and infrared radiation. Just as motion is variations in the straightness of strings, heat is variations in the straightness of strings within a bundle.

What we perceive as electromagnetic radiation is the stationary ripples that go out into space whenever a string shifts direction. These ripples can interfere with each other to result in a particular wavelength. The frequency of light, although not the wavelength, is an illusion due to the motion of our consciousness along the bundle of strings making up our bodies. The fact that strings gives off no ripples when the strings are perfectly straight is perceived by us as a body at a temperature of absolute zero that gives off no radiation, although it will reflect radiation from another source.

Our familiar Planck's Constant still gives the relationship between quantum energy and frequency in four dimension, just as it does in three, but the frequency is due to the movement of our consciousness. The ripples forming the radiation are actually stationary.

In our examples of the boat moving down the river or the car moving down the highway, the motion of the vessel caused trees to appear to move either together or apart. We knew that this was only apparent motion caused by the motion of the vessel and that this apparent motion of stationary trees could never seem to us to be faster than the boat or car was traveling. The apparent relative motion of two trees near each other would, in fact, be zero if the trees were directly ahead of or behind the boat or car and at one times the speed of the vessel if the trees were at right angles to the direction the vessel was traveling. In other words, the relative motion of stationary objects due to the motion of the vessel is equal to the sine of the angle between the vessel's motion and the stationary objects, such as trees.

Given this, you may be wondering why the apparent speed of light is always the same to us. If we look at objects at different angles, why would not the "speed of light" vary depending on the angle of the object, such as a star, relative to the direction that our consciousness is moving along the strings composing our bodies? In the same way that relative motion between two trees due to the motion of a car or boat varies according to the angle of the trees to the vessel's motion.

The reason that it does not vary is that whatever we look at in our universe, we are always looking toward the perpendicular relative to the direction that our consciousness is progressing along the strings composing our bodies and brains. Remember that we see only a one-dimensional slit of the fourth dimension, moving as our consciousness moves along the strings. This slit is always at right angles to the direction our consciousness is moving. That is why we always experience the speed of light at the maximum value, 186,282 miles per second, since the sine of ninety degrees is one. The "speed of light" seems to us to be one times the speed our consciousness is progressing along the strings composing our bodies and brains.

If we were able to see in the fourth dimension as well, not only could we see into the "past" and the "future", the speed of light would be at a maximum only at right angles to us in the fourth dimension and would seem to decrease until it was at zero at angles parallel to the strings composing our bodies, in other words the time or matter dimension. So, we would be unable to see directly into the past of future even if we could see in four dimensions instead of three.

Now that we can see that electromagnetic radiation such as light is only stationary ripples in space, it should become obvious why we can only see at right angles to the direction of travel of our consciousness. We cannot see into the past because our consciousness is moving away from the stationary ripples we see as light. We cannot see into the future because we have not yet arrived at the stationary ripples that we call light there yet. We are limited to experiencing and seeing three of the four dimensions simply because light consists of stationary ripples in the fabric of space while our consciousness is in motion along the bundle of strings composing our bodies.

POSTULATE 12—THE REASON WE CAN ONLY SEE AT RIGHT ANGLES TO THE DIRECTION THAT THE BUNDLES OF STRINGS COMPOSING OUR BODIES IS ALIGNED IS THAT LIGHT (AND OTHER ELECTROMAGNETIC RADIATION) IS STATIONARY RIPPLES IN THE FABRIC OF SPACE. WE CANNOT SEE THE PAST BECAUSE OUR CONSCIOUSNESS IS MOVING AWAY FROM THE STATIONARY RIPPLES IN THAT DIRECTION. WE CANNOT SEE THE FUTURE BECAUSE OUR CONSCIOUSNESS HAS NOT YET MET THE STATIONARY RIPPLES THERE.

If light were actually moving through space, instead of just our consciousness, we would be able to see into the future and would perceive four dimensions instead of three. If light was moving, but slower than our consciousness, we would still not be able to see into the past. However, if it was outpacing our consciousness we would be able to see into the past as well as the future. We would thus naturally see objects, and our bodies, as the bundles of strings that they are.

Since our consciousness is rushing toward the future, this would mean that if light was actually moving through space we would see the past in slow motion because the light would be delayed in "catching up" to our consciousness and the future would appear in fast forward because our consciousness would be rushing into light from that direction, kind of a variation of the Doppler Effect although it would vary according to the angle that we were looking into the fourth dimension. In that bizarre situation, we would see in four dimensions you could watch what was going to happen speeded up in one direction in space, then experience the event and then watch what had happened in slow motion by looking in the opposite direction. The other three dimensions would be the present as it is now. Although light would seem to be traveling faster than we usually perceive it due to the addition of the velocity of light, whatever it might be, to the velocity of our consciousness along the fourth dimension.

But since that is not what we see and we see only three dimensions, the present, we know that light consists of stationary ripples in space and it is only our consciousness that is moving.

Since you now understand why we see in only the three dimensions that we do instead of four, it may be obvious to you why we see strings as particles instead of strings.

Suppose a scientist has a microscope powerful enough to observe an electron. The electron is actually a string. However, the scientist's consciousness is moving along the strings composing his body and brain while light, or any other radiation from the electron, is stationary ripples in space. Thus, the scientist has moved past the light from the electron's "past" and has not yet gotten to the stationary ripples from the electron's "future".

All that the scientist sees is the light from the electron string that is at right angles to the dimension along which his strings are aligned. In a line in four dimensional space, there are three dimensions at right angles so, the scientist sees the electron string as a spherical particle in three dimensions instead of a string in four dimensions. (Yes, I know that an electron is much smaller than the wavelengths of light an so cannot be directly visibly observed, but this is just for an illustration.) You see, this theory is very simple yet it fits so many things together so perfectly.

QUANTUM PHYSICS

What we know as Heisenberg's Uncertainty Principle is caused by gradual shifts in the strings as well as the nature of the strings. Strings are mostly straight on a large scale but at quantum levels, this is not so. It seems to be rather like looking down at the ocean from an airplane. From 35,000 feet or so, the surface of the ocean appears to be perfectly smooth. However, from a ship it is a different story altogether. Gravity keeps the ocean surface smooth on a large scale. But there can be rough seas and waves if we take a closer look.

This is why strings, representing fundamental particles such as electrons and quarks, do not behave by the same standards as in "classical" or larger scale physics. Quantum particles are unpredictable just as the ocean surface is smooth on a large scale but not on a much smaller scale.

FORCE STRINGS

Aside from the strings we have discussed so far, there are other strings that are perpendicular to our matter strings and hold these strings together in bundles. These other strings are perceived by us to have energy but low mass or none at all. These are our so-called force particles or bosons. Force carriers come from perpendicular dimensions just as anything involving energy comes from a perpendicular dimension, one of the dimensions we perceive as space rather than the one we perceive as time. It requires energy to put together or separate matter strings. Therefore, it must come from one of the perpendicular dimensions.

Bosons are strings interacting with fermions (matter strings) but from a perpendicular dimension. Bosons appear to us as having positive ground state energies and tie the strings of matter particles together. Bosons naturally appear to us as points of energy. Bosons can be compared to twist ties or to the wires support-ing a drop ceiling. When a boson or other particle appears to us to exist for only a fraction of a second, it is due to the directional orientation of the string relative to our passing consciousness. A one-dimensional string perpendicular to us will seem to last only an instant.

We perceive both matter and force particles as infinitesimal points. We are, in our moving three-dimensional slice, seeing one point along a string for fermions and the end of the string for bosons. Force particles, the bosons, appear to have energy but brief existences due to spatial orientation. If bosons appear to us in our three dimensions to last briefly or move, it is simply due to the angle of the strings to our matter strings.

If gravitons, the particles carrying gravity, exist they would be force strings over great distances between bundles of matter strings. However, I believe gravity to be not actually a force requiring perpendicular strings but rather an innate property of space itself, which can be explained by the gravitational trenches in space. This is also in harmony with Einstein's General Theory of Relativity.

POSTULATE 13—IN FOUR-DIMENSIONAL SPACE, GRAVITY IS TRENCHES IN SPACE CAUSED BY THE MASS OF BUNDLES OF STRINGS AND PARALLEL TO THE BUNDLES. FORCE PARTICLES ARE STRINGS PERPENDICULAR TO AND TYING MATTER STRINGS TOGETHER.

CHAPTER 7

▼

MATTER IN SPACE

THE NATURE OF DIMENSIONS, MATTER AND IT'S STRINGS, ELECTRICAL CHARGES AND, MASS

We already can see what the "stationary" means in The Theory of Stationary Space. Now we come to the "space" part. We have been discussing strings that are of great length but infinitesimal cross-section. In other words, the strings are one-dimensional strings in four-dimensional space and aligned in one direction. This dimension we perceive as time and the other three we perceive as space.

Everything, except what is moved by living things, is essentially still. It is our consciousness, rushing past at the speed we perceive as the speed of light that makes the universe seem such an energetic place. When a string or a bundle of strings is slightly out of alignment in the fourth dimension, we perceive it as motion as our consciousness rushes past it. So complex is the processes required in our bodies and brains to create consciousness that we must rush through 186,282 miles of stationary, but intricately twisted, strings in our bodies and brains to get a second of consciousness. This is the essence of this theory.

But what exactly are these very long one-dimensional strings that make up what we perceive as matter? That is where the "space" part of the theory comes in. Considering what we know so far, there is a very simple and logical explanation for the existence of matter. We could consider this as "part two" of this theory.

No, it is not condensed or frozen energy as is commonly believed. This theory shows that energy does not really exist as we think of it. Energy is really strings of matter out of perfect alignment with the fourth dimension. So as our consciousness rushes past, we perceive it as motion because we can only see in three of the four dimensions. If this theory is correct, then there must be another explanation for the existence of matter.

The key to matter lies in space. The unit of space is the dimension. The most basic definition of space is the number of dimensions in that space. How many times can you turn a straight edge at a right angle in that space without going over the same position again? That would be the number of dimensions.

Obviously, dimensions can fit together to produce space of more than one dimension. My theory is that a dimension can "recognize itself" and a dimension cannot be twisted to make two dimensions out of one. The number of dimensions is conserved. Space can obviously bend, space is warped by massive objects to produce what we perceive as gravity. But dimensions cannot be twisted around to make more dimensions than there were to begin with.

POSTULATE 14—THE NUMBER OF DIMENSIONS IN A BLOCK OF SPACE IS CONSERVED. ONE OR MORE DIMENSIONS MAY NOT BE TWISTED TO PRODUCE MORE DIMENSIONS THAN THERE WERE ORIGINALLY.

My explanation for the existence of matter is that aside from the standard four dimensions, there are two contiguous dimensions that are not contiguous with the first four and are "folded" relative to the layout of the four. These two dimensions are reflected in the fact that there are two electrical charges in the universe, positive and negative.

Strings are not really one-dimensional strings but rather one-dimensional folds in a block of two-dimensional space. It may well be that it is this folding in the two-dimensional block that prevents it from being contiguous with the four-dimensional block of space. If the two blocks of space were contiguous, all we would have would be six dimensions of absolutely empty space. Instead due to the folding, the two-dimensional block shows up as one-dimensional strings along the folds in it. This is what we know as matter.

I find this to be a far more plausible explanation of the origin of strings than some of the other theories that I have heard such as the strings forming along "boundaries" of areas of space with different characteristics in the early universe.

Imagine these two matter dimensions as a flat sheet of paper. Let's start from the lower left corner of the paper. The bottom edge of the paper, or the width dimension, represents negative electrical charge. An electron has a negative charge and would be represented by a line in this dimension. We could also call this the x-axis. A particle with a positive charge would be represented by the y-axis. This would be along the height of the paper. We could say that one dimension is positive and one negative.

This, in fact, is our eighth great mystery solved; why matter exists at all, why is there not just space? The first seven mysteries were: What is time? Why does Newton's Law of Inertia always apply? Why does every action cause an equal and opposite reaction? Why are meaningful examples of entropy found only in living things? Why is the speed of light what it is instead of some other speed? Why would travel at the speed of light have any effect on the passage of time or the mass of an object? Why is the speed of light related to the conversion of matter and energy?

Yet another mystery that has always been baffling in physics is the exact origin of electric charges. Obviously, we know that there are two charges in the universe, negative and positive. But why is this so? Why are there two charges and not more? How exactly do the charges originate?

This is the answer, one charge for each of the two dimensions of space composing matter. This can also be seen in the electromagnetism of light and other so-called "electromagnetic" waves traveling through space. This is because whenever two dimensions fit together the "edge" of one takes on a positive charge and the other a negative charge like the mortise and tenon of a carpenter's joints. This holds the dimensions of space together.

When waves travel through multi-dimensional space, this underlying electromagnetism in space is exposed by the action of the waves. Thus, we perceive, and refer to, waves such as light as "electromagnetic". This is what gives away the fact that the electric charges actually originate with the dimensions.

The electromagnetism is actually a part of the underlying space and is merely "exposed" by the waves, although we know waves to be stationary ripples. In all electromagnetic waves, it is not the waves that are electromagnetic. The waves merely expose the electromagnetism of the space in which they travel. Space free of waves or ripples will not manifest electromagnetism because the two charges balance out unless the space is disturbed by waves. Gravity warps space on a larger scale without revealing the underlying electric charges.

This is most likely how a dimension of space would "recognize" itself so that it could not be twisted to form more dimensions than there were originally. No matter how many dimensions there were, only two charges would be necessary just as in mortise and tenon joints, only two types of edges are necessary.

Everyday matter is composed of protons and neutrons, as well as electrons. Protons and neutrons are composed of three quarks. These quarks are another fundamental particle composing strings. Protons and neutrons are made up of so-called "up" and "down" quarks. According to Murray Gell-Mann, the discoverer of quarks, a down quark has an electrical charge of -1/3. An up quark has an electrical charge of +2/3. Two up quarks together with one down quark have a charge of +1 and form a proton. Two down quarks and one up quark have a neutral electrical charge and form a neutron. Composite particles such as these and formed from quarks are known as hadrons.

There are other quarks and particles that are also represented by folds in the two-dimensional block of space. But if all matter particles except these disappeared tomorrow, probably only physicists would notice. Our everyday world is made up of combinations of these three particles, the up and down quarks and electrons. Contrary to the opinion of some physicists, different particles are not manifested by different vibrations in the same kind of string but the strings we perceive as particles are actually different due to the fold in the two-dimensional block of space of which they are composed.

On our sheet of paper representing the two matter dimensions, we have the x-axis representing negative charge and the y-axis representing positive charge. We know that these designations, negative and positive, are arbitrary and could just as easily be reversed.

Suppose we draw a line on the paper representing the matter dimension from the origin at 26.56 degrees? At this angle, the cosine of the angle is twice the sine. Since the cosine measures the x-axis, this would give us a charge of -1/3, just what we need for a down quark, just as an electron is represented by the x-axis. This is because if the cosine (representing negative) were twice the sine (representing positive), half of the negative charge would be spent canceling out the positive charge, leaving a charge of -1/3.

Now, let's draw a line at 80.54 degrees. We find that the sine of this angle is six times the cosine. After canceling, this would give us a charge of +2/3 for the up quark. This is why quarks are exceedingly difficult to pull apart. They are composed of two contiguous dimensions of space.

Now suppose we go to the top left corner of the paper and repeat our setup of lines just as described above except this time going downward instead of upward so that congruent lines from the two origins cross at right angles. The two strings crossing at right angles would have equal but opposite electrical charges. The line of an up quark, for example, would be perpendicular to the line of an anti-up quark. We could call this the anti-matter origin just as our first origin point gives us matter and this is just how matter and anti-matter operate in The Theory of Stationary Space.

We could also utilize color-coding just as is done in Quark Chromodynamics, the negative dimension could be one color and the positive another color and the strings formed by the folds would be represented by various shades of a mix of the two colors, such as white and black forming shades of gray for various strings representing quarks and anti-quarks.

When two strings meet in the four spatial dimensions that cross at right angles in the matter dimensions, there is no "common ground" at those angles and the two appear to us in our three dimensions to mutually annihilate. Keep in mind that this perpendicular crossing is the fold in the two-dimensional block of space that we perceive as matter and has nothing to do with how the particle and it's anti-particle, or string, meet in four-dimensional space.

POSTULATE 15—MATTER AND ANTI-MATTER CONSISTS OF TWO CONTIGUOUS DIMENSIONS OF SPACE, A NEGATIVE AND A POSITIVE DIMENSION. THIS IS THE ORIGIN OF ELECTRIC CHARGES. THESE TWO DIMENSIONS ARE FOLDED RELATIVE TO THE FOUR CONTIGUOUS DIMENSIONS OF SPACE. STRINGS CONSIST OF FOLDS IN THE TWO MATTER DIMENSIONS. THE SAME CHARGES IN THE FOUR CONTIGUOUS DIMENSIONS OF SPACE IS RESPONSIBLE FOR THE ELECTROMAGNETISM OF THE WAVES IN SPACE.

Have you ever wondered why a positron, which is a positively charged electron or the antimatter equivalent of an electron, will suffer mutual annihilation when it meets an electron? A proton has exactly the same charge as a positron yet protons combine with electrons to form atoms instead of mutually annihilating.

The reason is that the quarks in the proton have "common ground" with the electron in terms of the cosines of the angles, they have at least some of the x-axis in common, of their strings from the matter origin but the positron and electron have no such common ground in the two dimensions that from matter. The folds in the two-dimensional block of space forming electrons and positrons cross at right angles. However, the folds of the quarks are neither completely positive nor completely negative. The quarks are known to have charges of 2/3 and, -1/3 respectively and that provides common ground with the entirely negative electrons so that there are atoms of matter rather than mutual annihilation. Without this, matter as we know it would be unable to exist.

Basically, when strings meet that are formed by folds in the two-dimensional block of space that are at right angles to each other, they mutually annihilate. If the two strings are formed of parallel folds, they have the same charge and mutually repel. At angles in between, there is common ground to be found. If the two-dimensional block of space we know as matter was three-dimensional, there would possibly be a third type of matter.

The so-called strong nuclear force that overwhelms the electromagnetic force at very close distances to bind together the protons in the nucleus of an atom may consist of "common ground" provided by the neutrons in the nucleus. Notice that protons are never found together in nuclei without the presence of neutrons.

The reason that like electrical charges repel while unlike charges attract is the simple rules concerning dimensions of space. The electrical charges are composed of the two different dimensions of this matter space. When we bring two unlike charges together in our four spatial dimensions we are actually bringing the two dimensions together and since they are contiguous, they naturally attract each other. However when we bring two like charges together, we are violating the rule of conservation of dimensions and trying to bend a dimension around on itself and thus making more than one dimension out of it.

As we already know, this mutual repulsion of like charges is why two quantities of matter do not merge into each other. When you place a cup on a table, both objects are mostly empty space but the two remain separate and do not merge into each other due to the mutual repulsion of the negative electrons in the atomic orbitals of the atoms in the cup and those in the atoms of the table. This is the only reason that two objects cannot occupy the same place at the same time.

The four spatial dimensions are contiguous and the two spatial dimensions that we perceive as matter are contiguous but the two are not contiguous to each other. The characteristics of one block of space do not extend to the other. This is why matter is visible to us, the ripples in space that we know as light and other electromagnetic waves bounce off matter instead of passing right on through it. This is true even though "matter" is actually only space also.

Furthermore, the two spatial dimensions that we perceive as matter is folded relative to the four spatial dimensions. The folds are along those lines representing the particles as described above and it may, in fact, be these folds that prevent the two blocks of space from becoming contiguous. We have seen how space is capable of holding the gravitational warps and many, many different frequencies, amplitudes and, directional orientations of ripples at once. The two-dimensional block of space that we perceive as matter can likewise hold these folds, one for each and every string.

Imagine a piece of construction paper lying on a desk. This represents the four spatial dimensions. Now fold another piece of paper like a fan with the folds at the angles given above and also at every angle representing a string that exists and it's electrical charge components. Now place the folded piece of paper on top of the first piece of construction paper so that the only contact between the two pieces of paper is along the folds in the second, folded piece of paper. The folds would be at the angles we just discussed.

The flat piece of paper represents the four dimensions of space, including the one we imagine is time. The folded piece of paper represents the two dimensions of space that we perceive as matter. The two dimensions in this paper we perceive as the two electrical charges, positive and negative. Each fold in the paper we perceive as a different type of fundamental particle (which, of course, are strings in four dimensions). That, plainly and simply, is how the universe is constructed.

This has got to be the simplest and best explanation of the universe. This is it, the theory of everything. This is a plan for the universe simple enough to be stenciled on a T-shirt.

The two blocks of space are not mutually contiguous, so the four-dimensional block "makes room for" the two-dimensional block. The resulting warping of the four-dimensional block is what we perceive as gravity, similar in concept to the displacement of water by a ship. The ship and the water are not mutually contiguous but the water "makes room for" the ship. The two-dimensional block, of course, only exists in the four dimensions as one-dimensional strings along the folds in the two-dimensional block.

However the strings we experience as matter must be from a two-dimensional block, rather than from one dimension. This is the only practical way to explain the existence of two electric charges in the universe as well of the varieties of particles and anti-particles that can come from the two-dimensional block. If the strings were from one-dimensional space, there would have to be a different dimension for each type of fundamental particle and there would either be a different electrical charge for each particle or only one charge. The universe as we know it would be impossible. We cannot enter the two-dimensional block of space that we perceive as matter simply because we cannot, due to our nature, exist in fewer than four dimensions.

Incidentally, this scenario also explains two more great mysteries about the universe, numbers ten and eleven. We now see where mass comes from. Matter has mass, in other words it is acted upon by gravity or we could say that it warps four-dimensional space. However, some things, such as light and space itself, do not have mass. The answer is now simple and obvious. Anything composed of the two-dimensional block will have mass. Anything composed of the four-dimensional block will not have mass because this block will not logically warp itself.

Mystery number eleven is why is the universe so isotropic over great distances? That is, a large chunk of the universe is pretty much the same as any other large chunk of the universe. A theory called Inflation has been in use since the early eighties to explain this. If the universe, just after forming, suddenly expanded very rapidly then the matter composing it would not have had time to coagulate before the expansion placed it too far apart to do so.

However, if we consider the "Sandwich Model" of a block of two-dimensional space in a block of four-dimensional space much like a slice of cheese between two pieces of bread, this would account for the large-scale uniformity of the universe and inflation would be unnecessary for this purpose. This also helps account for why the shape of space in the universe is apparently flat, as opposed to spherical so that light returns to it's starting point or saddle-shaped.

POSTULATE 16—MATTER SEEM TO US TO HAVE MASS. HOWEVER SOME THINGS, SUCH AS LIGHT AND SPACE ITSELF, APPEAR TO BE MASSLESS. ANYTHING COMPOSED OF THE TWO-DIMENSIONAL BLOCK OF SPACE THAT WE EXPERIENCE AS MATTER WILL HAVE MASS. ANYTHING COMPOSED OF THE FOUR-DIMENSIONAL BLOCK OF SPACE WILL BE MASSLESS.

THE APPARENT EXPANSION OF THE UNIVERSE

As our consciousness progresses along the bundles of strings composing our bodies at 186,282 miles per second, astronomers see the universe as expanding. They have abundant evidence that galaxies are moving apart from each other and the further apart they are, the faster they are traveling apart.

What I am certain is actually happening is that the universe is not expanding at all. Everything is very still and placid. What is happening is that the universe is radial in structure like the spokes in a wheel and although the bundles of strings composing matter are aligned mostly in one direction, very distant objects such as galaxies are moving at an angle to us causing us to see these galaxies as rushing away from us because we can only see in three of the four dimensions. So actually, it is two dimensions that matter is aligned in but only on a very large scale. On any local scale, it is only one out of the four dimensions.

In a spherical universe, radial lines at a distance appear to be moving away and the red shifting of light or the Doppler Effect is due to the resulting angle that we intersect with the ripples in space (light) from those galaxies. By spherical, I mean the general shape of the universe and not any relativistic curvature of space. At the angle the ripples are spread by the angle over a wider distance and thus appear stretched out or redder. This is because red is the color of visible light with the lowest frequency. This red shift can be calculated using the Pythagorean Theorem.

The assumption ever since Edwin Hubble discovered the red shifting of the light from distant galaxies has been that this means the universe is expanding. What if we have been wrong ever since then? If there is no real motion, then obviously the universe cannot be expanding. We could say that the universe has already expanded.

There are indeed some astronomers who believe that there may be another explanation for the red shifting. Many ideas, from inflation to the speeding up of the cosmic expansion "discovered" in 1998 would have to be discarded or reevaluated. There is only the apparent expansion of the universe because our consciousness is moving along the strings composing our bodies outward from the center along the radial pattern of the universe.

When the universe was laid down, it does not seem as if it expanded at all. The two blocks of space were put together and the universe existed at once, although we can think of this point as the "Big Bang" if we wish. It could probably be described as a "sandwich" of the two-dimensional block in the four dimensions. The Big Bang was gravity, and possibly other factors, aligning the strings more or less in one dimension. Even though strings are formed from folds at angles in the two-dimensional block, they are aligned mostly in parallel in the four-dimensional block.

Astronomers today can detect the Cosmic Microwave Background radiation from the Big Bang. However, now that we are dealing with four spatial dimensions, this must be modified. If the universe is radial in form, as the "expanding universe" and red shifting of light seems to insists it must be, it is possible that the CMB radiation cannot be actually coming from the point of the Big Bang because then it would be parallel to the plane in which our perceived time and matter goes and would thus not be detectable by us, except for reflections or from a receiver traveling at high speed because this would send it out into a perpendicular dimension. Such radiation would be coming to us primarily from our past, not from a perpendicular dimension.

Since we can detect this radiation from any point in the sky, it is more logical to presume it is from all around when gravity aligned the strings mostly in their present dimension. Although somewhat more of it may have originated from the hub or origin area of the radial plan of the universe.

Another possibility worth considering is that the reason the universe appears to be constructed to a radial plan is that the folds composing the particles making up matter in the two-dimensional block of space radiate from the "origin" as we saw earlier, the x-axis representing electrons in one direction and the y-axis representing positrons in a perpendicular direction, both from the origin.

What about God? There is just too much evidence in favor of the Bible to leave that out. Where exactly is God? Suppose we add another dimension. That would make it seven altogether. The universe would look to God as a two-dimensional rug would appear to us with many intricate threads. God and heaven could be right here, right next to us but occupying more dimensions than we do much as an imaginary two-dimensional creature could not perceive a creature in three dimensions. This may be why Jesus said that "the kingdom of God is very near to you" even when it was nowhere to be seen. Our spirits are, in some way, able to access this dimension.

This may well be why the Bible contains such perfectly accurate prophecies. God can see the past, present and, future at a glance while mortals see only the present. The Book of Revelation also foretells a point when time will no longer exist. Now we know that it actually only exists in our perception anyway so doing without it altogether is a definite possibility.

CHAPTER 8

▼

SUMMARY

Einstein once stated "relativity is so beautiful that it must be correct". I think this theory must be fundamentally correct not because it is beautiful but because it explains so much about the universe while remaining so simple.

Remember that the principle of Occam's Razor is that the simplest explanation is usually the best. That may not be true when dealing with human beings but it does seem to be true in physics. Since the universe is composed entirely of space, as we might expect it is the spatial branches of mathematics such as geometry and trigonometry that we mainly use in this theory to describe it.

Not only that, but this simple theory explains many primal mysteries about the universe in one neat and coherent package. What is time from a physics point of view and why is the speed of light exactly what it is? Why does the speed of light play bizarre tricks with mass and time and why on earth does it have anything to do with the conversion of mass and energy? Why does Newton's Law of Inertia always apply and why does every action cause an equal and opposite reaction? Why are there two and only two electric charges and where do they originate? Why do some things have mass while others do not? Why does matter exist at all? Why is there not just space?

What has happened is that we have been deceived by our senses and have trusted our own perceptions far too much. Our consciousness is the reason we see the universe the way we do. We see the universe the way we do not only because of the way it is but because of what we are.

We must remember that God designed our consciousness for such things as farming and building, not so much for trying to understand the physics of the cosmos. This does not mean that we cannot understand this or that God does not wish us to understand it. Just that if we are going to really understand reality, we must learn to think outside the box a little bit, in this case outside our usual perceptions.

The truth is that our three-dimensional slice of the universe that we perceive at any given moment appears to us as much more complex then it actually is. It would appear considerably simpler if we could see all the relevant dimensions at once. Equal amounts of information will naturally appear more complex if presented in fewer dimensions rather than in more dimensions.

At this point, the universe has six dimensions. Three we perceive as space. Two we perceive as matter and the other one we perceive as time. Since we can see only in three dimensions, we do not see reality at all like it actually is and thus I have used the word "perceive" more in this book than ever before in my life.

What makes everything so different with the Theory of Stationary Space is that there is not the gaping holes in physics and cosmology that there were without this theory. This is not to say we know everything about the universe. But our model of the universe is not like a piece of Swiss cheese with gaping holes that physicists have gotten into the habit of just ignoring. Now it is more like cheddar cheese, we still do not know everything but these gaping holes have been filled.

Now, instead of just knowing that time goes by one second every second, the speed of light is absolutely fixed at 186,282 miles per second, this speed squared is the function in Einstein's famous formula E=MC squared, time stretches out and mass increases as we approach this speed, there are two electric charges in the universe and, some things have mass and some do not, we know why these things are so.

I think that this is a very big difference. Once we accept the definition of time as the direction in four dimensions of space that the strings composing matter are primarily oriented, the answers to all the other mysteries just seem to fall into place.

With this theory we can indeed put "the universe on a T-shirt". This model that explains so much that was not explained before can be stenciled on a T-shirt as a background square of one color, representing the four dimensions of space. Inside this square would be a slightly smaller square of another color to represent the two dimensions of space that we perceive as matter.

The smaller square would have lines in it. One line along the bottom border of the square would be along the "negative dimension" of this square and would represent electrons. Two lines would be drawn from the lower left corner of the smaller square. One at 26.56 degrees would represent the down quark with a charge of -1/3. Another line at 80.54 degrees would represent the up quark with a charge of +2/3. There are believed to be four more quarks that can be added if desired but these others are not believed to be important to ordinary matter as we know it.

These lines would be repeated from another origin point at the upper left corner of the smaller square so that congruent lines from the two origins would be perpendicular. These lines would represent antimatter just as the lines from the first origin represent matter. So that the right border of the smaller square would represent a positron. The lines are, of course, folds in the two-dimensional space that show up in the four-dimensional space as strings in four dimensions or as our familiar particles in two dimensions.

This is indeed the universe reduced to a design that could be stenciled on a T-shirt and is a secure base for building a greater understanding of the universe. How can this theory not be true? It is such a good mathematical model of the universe and for that to be the case, it must have a very high degree of congruence to physical reality.

This theory, as you may have concluded by now, forms the perfect marriage of relativity and string theory. The fact that this interpretation of string theory can explain all the whys of relativity that were never satisfactorily explained before seems to me to really give it credibility.

It is true that a lot of what was believed is reduced to being "apparent" or "perceived". There is no longer time but only apparent time, the distance our consciousness has traveled along the bundle of strings composing our bodies. There is no longer motion but only apparent motion, except for that associated with living things. There is no more speed of light but rather the speed of consciousness in four dimensions, the apparent speed of light in three dimensions. There is no longer space and matter but only space, some of which is apparent matter to us. Space is really all there is. There is no real expansion of the universe but only the apparent expansion of the universe. There is no need for the present Inflation Model to explain the uniformity of the universe, this theory and it's "Sandwich Model" does so.

Everything seems so different all of a sudden. But by sacrificing our existing view of the universe we get so much in return. We have solved the great mysteries of the universe. We know what time is from a physical point of view. We know why the speed of light is what it is and why bizarre changes to time and matter are associated with it and why, multiplied by itself, it is the function in Einstein's formula for the conversion of matter and energy. We can explain why Newton's Laws of Motion always apply. We know why we can only see in three of the four dimensions. We now know why there are two electric charges in the universe and why waves in space manifest electromagnetism. We know the exact nature of matter and why it exists. We know why some things have mass and some do not.

All I did was set out to find out what time was, but when I did all the rest just fell into place. An earlier theory of mine, The Theory of Primes, explained how the universe functions, it's operational formula. This theory shows how it is structured and why we perceive it as we do.

If there is one thing that conclusively proves the truth of this theory, it is the formula E=MC squared. It is easily the most famous formula in the world. But stop and think, why does the speed of something, particularly something (light) that does not even have mass at all, be the function for the conversion of mass to energy? Have you ever stopped to consider how bizarre this is? It only makes sense if this model of the universe is accurate.

We perhaps have a better understanding of where God is, when he seems to be right along side us but some have difficulty fitting this into a materialistic view of the cosmos, apparently our spirits have the ability to access the dimension that God inhabits but our mortal bodies cannot access. We certainly have a better idea of how we fit into the grand scheme of things and how our perceptions distort the real universe and give us instead a distorted, limited three-dimensional "apparent universe".

Do not be too surprised that our perceptions of so many primal components of the universe we live in were distorted. After all, it was not until Einstein's General Theory of Relativity came along that we knew we had been wrong in our perception of something as fundamental as gravity. It did not actually exist as a force, but was the warp in the fabric of space by the mass of the earth that caused us to perceive it as a force. With this theory, we now realize that our perceptions of time and the speed of light were similarly wrong simply because we can only perceive three spatial dimensions.

It is my belief that we are much better off now. We should really start thinking of the universe in terms of four dimensions and strings instead of three dimensions and particles. This is a solid foundation for universal models. There is certainly more to this, for one thing I did not explain just what force particles are that have no mass (or apparent mass) and do not appear to be space-exclusionary like matter (or apparent matter) is.

APPENDIX A

▼

LIST OF POSTULATES

POSTULATE 1—THERE ARE FOUR SPATIAL DIMENSIONS. MATTER IS COMPOSED OF VERY LONG ONE-DIMENSIONAL STRINGS ALIGNED PRIMARILY IN ONE DIRECTION IN THE FOUR DIMENSIONAL SPACE. IT IS THIS DIMENSION THAT WE PERCEIVE AS TIME. THE OTHER THREE DIMENSIONS WE EXPERIENCE AS SPACE.

POSTULATE 2—WE CANNOT MOVE AT WILL IN TIME SIMPLY BECAUSE IT IS THE DIRECTIONAL ORIENTATION OF THE STRINGS COMPOSING OUR BODIES THAT DEFINES TIME.

POSTULATE 3—WHAT WE PERCEIVE AS THE SPEED OF LIGHT IS ACTUALLY THE RATE AT WHICH OUR CONSCIOUSNESS MOVES ALONG THE BUNDLE OF STRINGS COMPOSING OUR BODIES AND BRAINS.

POSTULATE 4—OUR CONSCIOUSNESS IS PROGRESSING ALONG THE STRINGS OF OUR BODIES AT WHAT WE PERCEIVE AS THE SPEED OF LIGHT. EVERYTHING ELSE IS AT REST. THIS IS WHY WE PERCEIVE THAT NOTHING IS ABLE TO MOVE FASTER THAN THE SPEED OF LIGHT.

POSTULATE 5—WE CAN SEE ONLY THAT PORTION OF THE UNIVERSE THAT IS PERPENDICULAR TO THE POINT ON THE STRINGS MAKING UP OUR BODIES WHERE OUR CURRENT PARTICLE OF

CONSCIOUSNESS IS LOCATED. SINCE OUR UNIVERSE HAS FOUR SPATIAL DIMENSIONS, THAT MEANS WE SEE IN THREE DIMENSIONS. THE INFINITESIMAL SLICE OF THE FOURTH DIMENSION WE SEE IS WHAT WE CALL THE PRESENT.

POSTULATE 6—WE KNOW THAT THE OBSERVER PLAYS AN IMPORTANT ROLE IN QUANTUM PHYSICS. THE OBSERVER ALSO PLAYS A VITAL ROLE IN COSMOLOGY. IT IS IMPOSSIBLE TO UNDERSTAND THE UNIVERSE WITHOUT UNDERSTANDING THE ROLE THAT OUR OBSERVATION PLAYS IN IT.

POSTULATE 7—THE FURTHER WE GET FROM OUR EVERYDAY WORLD THAT WE WERE DESIGNED FOR, EITHER TO THE REALM OF THE VERY LARGE OR TO THE REALM OF THE VERY SMALL, THE LESS WE CAN RELY ON OUR OWN OBSERVATIONS AT FACE VALUE REGARDLESS OF THE QUALITY OF OUR INSTRUMENTS.

POSTULATE 8—IF TIME DOES NOT REALLY EXIST THEN MOTION, WHICH IS A FUNCTION OF TIME, CANNOT REALLY EXIST EITHER. WHAT WE PERCEIVE AS MOTION IS VARIATIONS IN THE STRAIGHTNESS OF STRINGS OR GROUPS OF STRINGS AS OUR CONSCIOUSNESS PASSES BY.

POSTULATE 9—THE REASON THAT MEANINGFUL EXAMPLES OF ENTROPY ARE FOUND ONLY IN LIVING THINGS AND ITEMS MADE BY LIVING THINGS IS THAT LIVING THINGS COMPRISE THE ONLY NEW MOTION IN THE UNIVERSE AND THE CONSCIOUSNESS IN LIVING THINGS IS MOVING IN ONE DIRECTION ONLY.

POSTULATE 10—NEWTON'S LAWS OF MOTION CAN BE EXPLAINED BY THE STRAIGHTNESS OF THE BUNDLES OF STRINGS COMPOSING BODIES OF MATTER. THE LAW OF INERTIA BECAUSE FROM THE POINT OF VIEW OF A STRAIGHT LINE, ANOTHER STRAIGHT LINE WILL SEEM TO HAVE A CONSTANT RATE OF MOTION (OR STILLNESS) WHEN VIEWED FROM SUCCESSIVE POINTS ON THE ORIGINAL LINE TO CONGRUENT POINTS ON THE SECOND LINE. NEWTON'S LAW CONCERNING OPPOSITE AND EQUAL REACTIONS BECAUSE THE AVERAGE DIRECTION OF

THE STRINGS COMPOSING MATTER ALIGNED IN SPACE IS ALWAYS CONSERVED.

POSTULATE 11—STRINGS PARALLEL TO THE STRINGS COMPOSING OUR BODIES WILL BE PERCEIVED BY US AS BEING AT REST. THE STRINGS PERPENDICULAR TO THE STRINGS COMPOSING OUR BODIES WILL BE PERCEIVED BY US TO BE TRAVELING AT THE SPEED OF LIGHT.

THE APPARENT VELOCITY IN THREE DIMENSIONS IS GIVEN BY THE SINE OF THE ANGLE BETWEEN THE TWO MULTIPLIED BY THE APPARENT SPEED OF LIGHT. THE TIME DILATION IN THREE DIMENSIONS IS GIVEN BY THE COSINE OF THE ANGLE BETWEEN THE TWO.

POSTULATE 12—THE REASON WE CAN ONLY SEE AT RIGHT ANGLES TO THE DIRECTION THAT THE BUNDLES OF STRINGS COMPOSING OUR BODIES IS ALIGNED IS THAT LIGHT (AND OTHER ELECTROMAGNETIC RADIATION) IS STATIONARY RIPPLES IN THE FABRIC OF SPACE. WE CANNOT SEE THE PAST BECAUSE OUR CONSCIOUSNESS IS MOVING AWAY FROM THE STATIONARY RIPPLES IN THAT DIRECTION. WE CANNOT SEE THE FUTURE BECAUSE OUR CONSCIOUSNESS HAS NOT YET MET THE STATIONARY RIPPLES THERE.

POSTULATE 13—IN FOUR-DIMENSIONAL SPACE, GRAVITY IS TRENCHES IN SPACE CAUSED BY THE MASS OF BUNDLES OF STRINGS AND PARALLEL TO THOSE BUNDLES. FORCE PARTICLES ARE STRINGS PERPENDICULAR TO AND TYING MATTER STRINGS TOGETHER.

POSTULATE 14—THE NUMBER OF DIMENSIONS IN A BLOCK OF SPACE IS CONSERVED. ONE OR MORE DIMENSIONS MAY NOT BE TWISTED TO PRODUCE MORE DIMENSIONS THAN THERE WERE ORIGINALLY.

POSTULATE 15—MATTER AND ANTI-MATTER CONSISTS OF TWO CONTIGUOUS DIMENSIONS OF SPACE, A NEGATIVE AND A POSITIVE DIMENSION. THIS IS THE ORIGIN OF ELECTRIC CHARGES. THESE TWO DIMENSIONS ARE FOLDED RELATIVE TO THE FOUR

CONTIGUOUS DIMENSIONS OF SPACE. STRINGS CONSIST OF FOLDS IN THE TWO MATTER DIMENSIONS. THE SAME CHARGES IN THE FOUR CONTIGUOUS DIMENSIONS OF SPACE IS RESPONSIBLE FOR THE ELECTROMAGNETISM OF THE WAVES IN SPACE.

POSTULATE 16—MATTER SEEMS TO US TO HAVE MASS. HOWEVER SOME THINGS, SUCH AS LIGHT AND SPACE ITSELF, APPEAR TO BE MASSLESS. ANYTHING COMPOSED OF THE TWO-DIMENSIONAL BLOCK OF SPACE THAT WE EXPERIENCE AS MATTER WILL HAVE MASS. ANYTHING COMPOSED OF THE FOUR-DIMENSIONAL BLOCK OF SPACE WILL BE MASSLESS.

APPENDIX B

▼

GLOSSARY OF TERMS

ANTIMATTER—Matter with opposite charges to conventional matter. A positron orbits the nucleus instead of an electron. Matter and anti-matter undergo mutual annihilation upon contact. In the Theory of Stationary Space, antimatter is formed of folds in a two-dimensional block of space just as matter is but the folds forming the strings of antimatter are at right angles to their corresponding matter strings and thus there is no "common ground" upon meeting and mutual annihilation occurs. It is important to remember that matter and anti-matter is formed from folds in the same two-dimensional block of space but at right angles to each other.

APPARENT—We see the universe only in three dimension when there are actually four dimensions of space. Also time, motion and, the speed of light are tricks that our perception plays on us. The result is that the "apparent" universe that we see is not the same as the "real" universe in four dimensions would appear.

BOSON—A force string, as opposed to a matter string. Bosons are at right angles to matter strings and accomplish tasks such as holding together the protons in a nucleus.

BRANE—The abbreviation of membrane. The universe is believed by some to be a brane of three-dimensional space floating around in space with higher dimensions.

CHARGE—One of two opposite charges in the universe. We arbitrarily refer to these two charges as "negative" and "positive".

CLASSICAL PHYSICS—Physics prior to the introduction of relativity. Classical Physics is also referred to as "Newtonian Physics" and is concerned with everyday science revolving around rules that are familiar to most people.

COMMON GROUND—The factor that determines whether two strings (particles in our three dimensions) will undergo mutual annihilation upon meeting. If the strings of the two particles are perpendicular to each other on the two-dimensional block of space that we perceive as matter, they have no "common ground" and will undergo mutual annihilation. The two particles are usually referred to as a matter-antimatter pair.

COSINE—When a radius and the resulting angle is described in a quadrant or circle from the x-axis (horizontal plane) and the y-axis (vertical plane), the cosine of the angle of the radius is defined as X/R. This is the length of the x-axis within the range of the radius divided by the length of the radius. The cosine begins at one at zero degrees (the x-axis) and goes to zero at 90 degrees (the y-axis).

DIFFERENT POINTS MODEL—The theory that for two persons to talk or otherwise interact, they need not and are almost certainly not, having their present moments of consciousness perpendicular to each other, in other words at the same time. You, in your present moment of consciousness, are interacting with that person as they were or will be when they have their moment of consciousness perpendicular (in other words in the present) to where you are having your moment of consciousness as the interaction takes place. I prefer this model to the Frontier Model.

DIMENSION—These are the basic components of space. A simple line is one dimension, a piece of paper is two dimensions and we live in three dimensions. The number of times a straight edge can be turned at right angles without going over the same position twice. Space is made of one or more contiguous dimensions.

DOPPLER EFFECT—The increase in the wavelength of a wave when the source of the wave is moving toward the observer and the corresponding decrease when the source of the wave is moving away from the observer. Best illustrated by the sudden lowering of a train's whistle as the train passes a person standing by the track. In physics, it usually refers to the red shifting of the light from a galaxy due to it's apparent motion away from us at high speed.

DOWN QUARK—A quark is one of the fundamental components of matter. A down quark has an electrical charge of negative 1/3.

ELECTROMAGNETIC—A basic force in the universe involving the attraction of like charges and repulsion of unlike charges. Radiation such as light is also electromagnetic in nature.

FERMION—Particles composing matter that show up as strings in four dimensional space. Quarks and leptons, which include electrons, are fermions. In contrast to fermions, bosons are force particles.

FORCE—A fundamental governor of the universe. The forces are the strong nuclear force that holds the atomic nucleus together, the weak nuclear force associated with radioactivity and electromagnetism. Some people consider gravity as a force also but the General Theory of Relativity defines it as the warp of space due to a massive object.

FOURTH DIMENSION—The dimension which humans and other living things perceive as time. This dimension is actually another spatial dimension but is the dimension in which the bundle of strings composing our bodies and other strings composing matter are aligned. The movement of our consciousness at what we perceive to be the speed of light along these strings is what makes us perceive this dimension as time. This dimension represents matter while the other three represent energy.

FRONTIER MODEL—When two persons talk or otherwise interact, they naturally seem to be having their present moments of consciousness at the same time. This may be because all humans' moments of consciousness are coordinated into a kind of frontier or wave so that all living people are having their moments of consciousness perpendicular to each other, in other words at the same time. I prefer the Different Points Model to this one.

GRAVITY—The warp in the fabric of space due to a massive object such as a planet. It has the effect of pulling other objects toward, or in orbit around, the planet.

INFLATION—The model accepted by many cosmologists to explain the uniformity of the universe on a large scale. If the universe expanded very, very rapidly shortly after it's birth then the matter in the universe would be moved too far apart to coagulate when the expansion slowed back down. This Theory of Stationary Space instead offers the "Sandwich Model" of a block of two dimensional space in a four dimensional block to explain this uniformity of the cosmos.

LEPTON—Fundamental components of matter not composed of quarks and not subject to the strong nuclear force. The most important lepton is the electron.

M-THEORY—A relatively recent variation of string theory in which one-dimensional strings are replaced by two-dimensional membranes.

MASS—Anything that is acted upon by gravity, in other words warps the fabric of space with it's presence.

MATTER—That which has weight and takes up space and is composed of fermions. An example is this book.

MATTER BLOCK—The two-dimensional block of space that is "folded" relative to the four-dimensional "space block" of space and that is perceived by us as matter. There are two opposite electrical charges in the universe because the matter block has two dimensions. Particles (strings in four dimensions) consist of the folds in the matter block relative to the space block. Strings (particles in three dimensions) from electrons to quarks and anti-quarks to positrons have various charges determined by the angle of the fold in the matter block from which they are formed.

MEAN SPATIAL ALIGNMENT—The direction that the strings and bundles of strings composing what we know as matter is conserved. According to a law of Newton, every action causes an equal and opposite reaction. This means that some bundles of strings, representing objects of matter, will be out of parallel alignment but that the average overall parallel alignment will always remain the same.

MEMBRANE—A two-dimensional sheet in M-Theory in place of one-dimensional strings in conventional string theory.

NEW MOTION—Motion produced or caused by living things. Essentially, the only changes since the original laying down of the universe. This is why meaningful examples of entropy always involve living things or items made by living things. Since the consciousness of living things moves in one direction only, this causes time to appear irreversible due to entropy.

NEWTONIAN PHYSICS—Physics prior to the introduction of relativity. Classical Physics is also referred to as "Newtonian Physics" and is concerned with everyday science revolving around rules that are familiar to most people.

QUANTUM PHYSICS—The physics concerning the basic nature of the universe on a very small scale but in harmony with cosmology.

QUARK—One of the fundamental components of matter. There are believed to be six types of quarks. The only ones important to everyday matter are the up and down quarks.

RADIAL—A circular pattern like the spokes on a wheel.

RED SHIFT—The apparent result of the Doppler Effect. The shift to red of the light from distant galaxies believed to be moving away from us at high speed.

RELATIVITY—Two theories by Albert Einstein. The first, the Special Theory, concerns the invariableness of the speed of light but the variableness of time. The second, the General Theory, concerns the warping of space by massive objects to cause what we perceive as gravity.

RELATIVITY OF THE PRESENT—Time does not really exist except to our consciousness. That means that all of the past, present and, future exist at once in a fourth dimension that is actually space, not time. Our lives likewise exist all at once but we are experiencing only one moment at a time. This causes the perception that everyone we interact with is having their moments of consciousness concurrently with ours. The truth is that our moments of consciousness, while progressing in the same direction and at the same velocity, are almost certainly widely separated by distance. Whatever today's date is to you, it is probably not the same at the present moment in terms of consciousness for someone you may interact with. They will have, or have had, the interaction when they are at that date. You see them not as they are "now" but as they were or will be on that date.

SANDWICH MODEL—The model offered by this theory to explain the amazing uniformity of the universe on a large scale. The universe is a "sandwich" of a block of two-dimensional space in a block of four-dimensional space. This accounts for the uniformity and previous theories to explain this uniformity, such as inflation, are not necessary for this purpose.

SINE—If in a quadrant or a right angle there exists a radius line at some angle between the x-axis at zero degrees and the y-axis at 90 degrees, the sine of the angle of the radius is defined as the ratio y/r. The sine of an angle ranges from zero at zero degrees to one at 90 degrees because at 90 degrees, the radius line and the y-axis would be one and the same.

SPACE BLOCK—The four contiguous dimensions of space. These dimensions are perceived by living things as three dimensions of space and one dimension of time.

SPEED OF LIGHT—The speed at which light and other electromagnetic radiation appears to us to travel. The speed is 186,282 miles per second or 300,000,000 meters per second in a vacuum.

STRINGS—Fundamental particles such as quarks and electrons as seen in three dimensions are really believed to be strings in four dimensions.

SUPERSTRINGS—A series of theories, starting in 1970, that the fundamental particles of the universe are really strings of tiny cross-section in more dimensions than living things can experience.

TIME—The fourth dimension as perceived by humans and other living things. We cannot move in the fourth dimension at will because that is the direction in which the bundle of strings composing our bodies is aligned. We can only see a one-dimensional slit of this fourth dimension as our consciousness moves forward on the strings composing our bodies at what we perceive to be the speed of light. So, we perceive this dimension as time manifested as motion.

UP QUARK—A quark, a fundamental particle, with a charge of +2/3. Two up quarks and one down quark (-1/3 charge) form a proton with a charge of 1. Two down quarks and one up quark form a neutron with no charge.

X-AXIS—In a circle or quadrant, the horizontal line is usually referred to as the x-axis and is where the circle or quadrant begins, with an angle of zero.

Y-AXIS—In a circle or quadrant, the vertical line is usually referred to as the y-axis and represents 90 degrees in a quadrant. The x- and y-axes are perpendicular, or at right angles to each other.

0-595-33909-3

www.ingramcontent.com/pod-product-compliance
Lightning Source LLC
Chambersburg PA
CBHW021016180526
45163CB00005B/1982